The
PIPER
INDIANS

Bill Clarke

TAB BOOKS Inc.
Blue Ridge Summit, PA

FIRST EDITION

SECOND PRINTING

Copyright © 1988 by TAB BOOKS Inc.

Printed in the United States of America

Reproduction or publication of the content in any manner, without express permission of the publisher, is prohibited. No liability is assumed with respect to the use of the information herein.

Library of Congress Cataloging in Publication Data

Clarke, Bill (Charles W.)
    The Piper Indians.

    Includes index.
    1. Piper airplanes.   I. Title.
TL686.P5C57 1988   629.133′343   87-33509
ISBN 0-8306-0232-1
ISBN 0-8306-2432-5 (pbk.)

TAB BOOKS Inc. offers software for
sale. For information and a catalog,
please contact TAB Software Department,
Blue Ridge Summit, PA 17294-0850.

Questions regarding the content of this book
should be addressed to:

    Reader Inquiry Branch
    TAB BOOKS Inc.
    Blue Ridge Summit, PA 17294-0214

Cover photograph courtesy of Piper Aircraft Corp.

Contents

Acknowledgments

This book was made possible by the kind assistance and contributions of:

Ken Johnson of AVCO Lycoming
A.O.P.A.
Federal Aviation Administration
Joe Ponte and his staff at Piper Aircraft Company
Joe Christy

. . . and all those other wonderful "airplane" people who provided me with photographs, descriptions, advice, hardware, and friendship.

Introduction

In 1958, Piper Aircraft introduced the Comanche. This was the first of the single-engine all-metal Pipers, and was also the start of a new era for the company. Gone forever were the tube-and-fabric "Cub-type" airplanes.

Since then, many different models of all-metal planes have been built by Piper. Most have been named after famous American Indian tribes or common Indian objects.

The Piper Indians is intended to assist the pilot, the owner, the would-be owner, or the aviation buff in gaining a complete understanding of the entire series of all-metal Piper airplanes.

In this book you will learn about the various years/models and their differences, read about the chronic problems—and how to fix them, and learn about modifications that can be made to improve performance and comfort.

If you're thinking of purchasing a used Indian, you will discover where and how to locate a good one—and, more importantly, how to get the most for your money in the process. Although the prospective buyer may have a basic idea of what the advertised airplane looks like, he should have a source to review for further information about the airplane, its equipment, and value. This book is such a source.

A walk-through of the purchase paperwork will be taken, with examples of the forms shown. Also included is a price guide, based upon the current used airplane market.

Read how to care for your plane, and learn what preventive maintenance you may legally perform yourself. See what an annual inspection is all about. Improve your engine operations, while saving fuel and extending the life of this expensive part of the airplane. An avionics section is included to aid you in making practical/economical decisions should you decide to upgrade your instrument panel.

Hangar-fly the Indians and see what the pilots of these planes have to say. Hear from the mechanics who service them, and read what the National Transportation Safety Board has to say about the Indians (and other small family airplanes).

In summary, *The Piper Indians* was written to educate the owner/pilot by exposing him to as much background material as possible in one handy-sized reference guide.

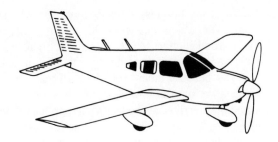

Chapter 1
Piper History

The mere words "Piper Cub" are often used in reference to any small airplane. Perhaps this is because there is no other name more associated with small airplanes than Piper. When I was young, there used to be several small airports near my home, and most of the local folks referred to all the small airplanes on these airports as "Piper Cubs."

EARLY HISTORY

The Piper Aircraft Corporation began life as the Taylor Brothers Aircraft Corporation, formed by C.G. Taylor. William T. Piper bought a few shares of the fledgling Taylor company as an investment when Taylor first moved his company to Bradford, Pennsylvania. Piper was not involved with the airplane manufacturing business at that time.

When the Taylor Brothers Aircraft Corporation failed—as many small companies did during the depression era—Piper bought the remaining assets for $600. He then formed the Taylor Aircraft Corporation and promptly gave half of the assets back to C.G. Taylor. (Never let it be said that Piper was a selfish man.)

About this time, Taylor designed a small two-place trainer called the Cub. The Cub was intended to be a cheaper, less-complex trainer than was currently on the market. The only similar aircraft at that time was the Aeronca.

All these business dealings took place during the Depression years. This was a time when, for most people, money was scarce for food and shelter. Jobs were scarce. Flying was for the rich. Life for a fledgling aircraft company was difficult at best.

The financial times were so hard that, at one point, airplane engines were available to Piper only on a cash-and-carry basis. When a customer arrived at the factory to purchase a new airplane, a Piper employee made a quick trip to the railroad express office to pick up an engine. All new engines were stored at the express office until paid for in cash. After the cash was paid, the engine was taken to the factory where it was installed on the new airplane. All this took place while the customer waited.

In 1935 the partnership with Taylor was dissolved. Taylor went to Alliance, Ohio, and formed the Taylorcraft Aviation Company. In 1937 the original Bradford airplane factory was destroyed by fire. After the fire, the factory was relocated to Lock Haven, Pennsylvania, and the name changed to Piper Aircraft Corporation.

Not long after Piper inaugurated business in Lock Haven, the clouds of World War II began forming. It was a recognized fact that the war would be a test of air superiority. The United States government, seeing the air race coming, shouted a mighty call for pilots. As a result, Piper Aircraft began producing large numbers of small trainer airplanes. The company states that four out of five World War II pilots received their original flight instruction in these small planes. They were called Piper Cubs.

Piper produced the L-4, a military version of the J-3 Cub, during the war. These planes were used for light transport during the war.

After the war, Piper Aircraft, like all the other airplane builders, produced large numbers of small planes to quench the postwar flying thirst. Unfortunately, in 1947 the sales boom went bust.

By late 1947, Piper was pared to 157 employees, down from 1,738. Recovery was slow, and only done with the most careful of planning and the introduction of practical airplanes.

By 1951, Piper had seen the introduction of the Vagabond (a "poor man's" airplane), the four-place Pacer, and the Tri-Pacer. Billed as easy to fly, the Tri-Pacer's sales took off.

In 1954 the first Piper twin-engined airplane, the Apache, was introduced. Profits from the Apache gave Piper the necessary funds to expand.

In the 1960s a new production plant was built for Piper in Vero Beach, Florida. In addition to aircraft production, considerable research was conducted at the new Florida facility—no doubt due to the better year-round flying conditions than those provided by the long and cold Pennsylvania winters.

Additional fabrication plants were built in Pennsylvania, but it was in Florida that Piper designed and built the larger Indian models that now serve the commuter airlines and corporate users. The Lock Haven plant is part of Piper history now, as are the planes once produced there.

History sometimes produces strange turns of events. One such point of interest is that Taylorcraft airplanes, of C.G. Taylor design, were being built until early 1987 in the old Piper plant at Lock Haven—the very plant that once produced the Piper Cub.

It's interesting that William T. Piper did not start his aviation career until he was nearly 50 years of age. Mr. Piper died in 1970, at the age of 89.

Piper Aircraft later became a division of Lear Siegler, Inc., a diversified manufacturer with operations in numerous fields including aerospace, electronics, automotive, agricultural, and recreation.

In 1987, Piper was sold to M. Stuart Millar, a private businessman.

CHRONOLOGY OF THE PIPER CLASSICS

The most famous word associated with Piper airplanes is "Cub." Cub actually was the name applied to several models of Taylor/Piper airplanes. The first Cub of Piper manufacture was the model J-3 (Fig. 1-1). A simpler flying machine has never been made.

A pre-World War II design airplane, the J-3 Cub was introduced in 1938. Civilian production was halted during the war years; however, the J-3 went to war as the Army L-4. The L-4 was used for liaison, VIP transportation, and tactical observation. Among the VIPs who rode in the L-4 were George Patton, Omar Bradley, Dwight Eisenhower, and Winston Churchill. At one time, J-3s were coming out of the factory door at the rate of one every 10 minutes.

Civilian production was restarted in 1945, and continued through 1947. A total of 14,125 J-3s were built. The J-3 was the basis for many other Piper models through the Super Cub (Fig. 1-2).

The J-4 Cub Coupe was built before the war, from 1939 to 1941. The Coupe used many J-3 parts, which kept costs of design and production to a minimum. No J-4s were built after the war. They seated two, side-by-side, and were powered with a 65-hp engine.

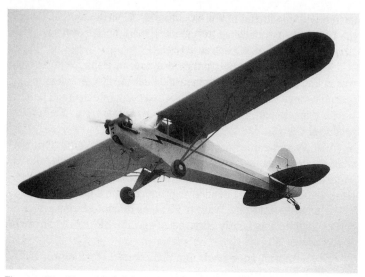

Fig. 1-1. The Piper J-3 Cub is the best-known Cub-series aircraft. A good J-3 will fetch more than $10,000 on today's market.

The J-5 Cub Cruiser was introduced just before the war, in 1941, and seated three persons. The pilot sat up front in a single bucket seat, and there was a bench seat for two to sit—very cozily—in the rear. Like the J-4, it shared many J-3 parts. After the war the J-5C,

Fig. 1-2. The Cub emblem, as found on all original (and most rebuilt) Cubs.

an improved version of the J-5, was brought out as the Super Cruiser. Improvement, in aviation, often means the same old airframe with a larger engine. The J-5C was powered with a 100-hp engine.

The J in the early Piper model numbers was for Walter C. Jamouneau, Piper's Chief Engineer. Later, the PA designator (for Piper Aircraft) replaced the J.

The PA-11 Cub Special was introduced in 1947 as a modernized version of the J-3. The improvements allowed for solo flight from the front seat and a fully cowled engine, available in either 65 hp or 90 hp. 1,428 Specials were built before production ceased in 1949.

The PA-12 Cruiser was an updated J-5. It still had the three-seat configuration, but added a slightly larger engine.

The first Piper production model of a four-place airplane was the PA-14 Family Cruiser. First built in 1948, the PA-14 is really a modified PA-12, with an extra seat up front. The Family Cruiser was small and underpowered. Only 237 were manufactured. The price for a new PA-14 was $3,825.

The PA-15 was introduced as the Vagabond in 1948. A very basic airplane, it had a Lycoming O-145 engine rated at 65 hp, and seated two side-by-side. The main landing gear were solid, with the only shock-absorbing action coming from the finesse of the pilot at landing time.

The PA-16 Clipper was introduced in 1949 as an update to the PA-14. The Clipper had shorter wings and fuselage than the PA-14, and was really the first of the ''short wing'' Pipers. The PA-14 and the PA-16 are unique from the standpoint of controls. They have sticks, rather than the control wheels usually associated with four-place airplanes.

A new, improved Vagabond came off the assembly line later in 1948. The new PA-17 Vagabond was a plusher airplane than the PA-15. It had bungee-type landing gear, floor mats, and a Continental A-65 engine. The Continental engine had more ''punch'' than the Lycoming, yet both were rated at 65 hp.

The Super Cub, PA-18, is basically an airplane whose roots can be traced back to the J-3. Although built with a completely redesigned airframe, the PA-18 looks like a J-3. First built in 1949, it's a fun plane to fly; however, it's usually found working for its keep. Super Cubs are utilized in photography, mapping, fish spotting, spraying, glider towing, bush flying, etc.

In 1950, the PA-20 Pacer series was begun, as an updated four-place plane. Like the PA-14s and -16s, the PA-20s were of tube-and-fabric construction and conventional gear design. A total of 1,120

Pacers were built. Evolution of the Pacer saw the eventual addition of a nosewheel, the latter resulting in the PA-22 Tri-Pacer.

Billed as an "anyone can fly it" airplane due to the tri-gear, the PA-22 was first sold as an option to the PA-20 Pacer. Sales of the nosewheeled Tri-Pacer soared, while those of the conventional-geared Pacer fell. As a result, the Pacer was removed from production in 1954.

The PA-22s were built from 1951 until 1960 and can be found with 125-, 135-, 150-, and 160-hp engines. A total of 7,670 Tri-Pacers were built before Piper moved on to all-metal four-place airplanes (Fig. 1-3).

Piper's last tube-and-fabric airplane was the PA-22-108 Colt, a trainer. It was really a two-seat, flapless version of the Tri-Pacer. The Colt was powered with a 108-hp Lycoming engine; 1,820 were built between 1961 and 1963.

CHRONOLOGY OF THE INDIANS

When Piper Aircraft introduced its first all-metal airplane, the tradition of utilizing American Indian tribal and weapon names for different aircraft models was adopted.

Apache: *A nomadic North American Indian tribe found in the southwestern area of the United States.* Piper's first airplane named for an American Indian tribe was a light twin called the Apache. First introduced in 1954, the PA-23 saw many changes over the years,

Fig. 1-3. The PA-22 Tri-Pacer was the last of the tube-and-fabric Piper aircraft. They are economical to purchase, but due to the fabric cover, can be expensive to refurbish.

Fig. 1-4. The all-metal PA-28 design was a complete departure from tube-and-fabric airplanes. The first PA-28s became the backbone for all later models of Piper airplanes.

eventually evolving into the Aztec. The last year of Apache production was 1963.

Comanche: *A tribe of North American Indians that once roamed the south Plains.* In 1958 the Comanche was introduced as an all-metal, single-engine, retractable-geared airplane. Available in various engine powers, the Comanche remained in production until 1972.

In 1963 Piper brought out a new light twin, the PA-30. Although somewhat similar in appearance to the Cessna 310, the Twin Comanche is much lighter. Powered with two Lycoming 160-hp engines, it sits low to the ground, and is very docile to fly. Production of the Twin Comanche ended in 1972.

1970 saw Piper improve the PA-30 by adding counter-rotating engines, thus eliminating the "critical engine" effect. At this time the model number changed to PA-39; however, the name Twin Comanche remained.

Aztec: *Central American Indians noted for their highly advanced civilization.* In 1960, the first sales of the Aztec PA-23 were made. It was powered with twin 250-hp Lycoming engines, and seated five. After 1961, all Aztec models had six seats. The Aztec shares the PA-23 model number with the Apache.

Cherokee: *A tribe of North American Indians that once dwelled in the North Carolina and Tennessee area.* In 1961 Piper launched the PA-28 Cherokee series. The PA-28 series was destined to be the backbone of all Piper single-engine airplanes (Fig. 1-4).

Improvements would come in the form of larger engines, modified wings, stretched cabins, and retractable landing gear.

Initially, the PA-28 was powered with a 150-hp Lycoming engine. It was called the PA-28-150. An optional version, the PA-28-160 with a 160-hp engine, was also available.

The Cherokee 140 was introduced in 1964 as a two-to-four-place trainer. This was the first training aircraft Piper produced since the end of Colt production. In reality, the 140 was a standard PA-28 without the rear seat; otherwise it was the same airframe as a four-place PA-28.

1965 saw the introduction of a much larger Cherokee, the PA-32. Called the Cherokee Six, it was available with 260-hp or 300-hp engines. Equipped with removable seats, this craft can carry cargo, a stretcher, or even livestock.

Arrow: *A straight, thin shaft, shot from a bow, known for its speed and accuracy.* In 1967, the PA-28R Arrow, with retractable landing gear, was ushered in. Simplicity is the key to the operation of the Arrow. Even the responsibility of controlling the landing gear was removed from the pilot—it's automatic!

Seneca: *A tribe of North American Indians once found in the Finger Lakes region of western New York.* In 1971, Piper introduced a new light twin called the Seneca. It was basically a Cherokee Six made twin. Called the PA-34, the Seneca is a very roomy light twin.

Warrior: *An Indian experienced in battle.* In 1974, the Warrior series was brought out. The Warriors were the PA-28 with longer,

Fig. 1-5. The Tomahawk represents a "state-of-the-art" modern trainer.

redesigned wings and a stretched fuselage. Piper claimed, "The Warriors are a logical progression of the PA-28 series."

Performance changes resulted in expanded carrying abilities and extremely gentle stall characteristics. The new wing, often called the "Warrior wing," was later added to all PA-28-series airplanes.

Tomahawk: *A light axe used as a weapon or tool by many North American Indians.* In 1978, the PA-38 Tomahawk was introduced as a trainer. The Tomahawk, an all-metal low-wing airplane, bears no resemblance to the earlier tube-and-fabric Piper trainers. It was the first true two-place trainer built by Piper since Colt production was halted in 1963 (Fig. 1-5).

Members of the Piper tribe used for corporate and commuter-airline operations are the Aerostar, Mojave, Navajo, Chieftain, and Cheyenne. The Pawnee and Pawnee Brave are used for chemical application (crop dusting).

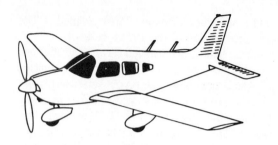

Chapter 2

The Original PA-28 Cherokees

When Piper Aircraft Corporation introduced the PA-28 in 1962, it not only initiated a new product line, it signaled the end of an era.

The valiant airplanes of tube-and-fabric construction had come and gone. Now, using new construction techniques, the Vero Beach facility was producing an all-metal, monocoque-design airplane. Modern times had arrived.

The original PA-28 Cherokee series was designed by John Thorpe (of EAA homebuilt fame) and Fred Weick (of Ercoupe fame) to be the ultimate in operational simplicity, and of low-cost manufacture. The one-piece stabilator design, patented by John Thorpe, and a hefty tricycle landing gear system are examples of this planned simplicity. The wings are of laminar-flow design, only 30 feet long, and quite thick. They are often referred to as "Hershey bar" wings. There were approximately 400 fewer parts used in the construction of the PA-28 than in the PA-22 Tri-Pacer.

The overall airplane design was a master stroke, as the PA-28 became the design backbone for almost all the subsequent models that Piper would build (Figs. 2-1 through 2-5).

First introduced with a 150-hp Lycoming engine, models soon became available with 160- and 180-hp engines. In 1964 the PA-28-235 was introduced, flexing the muscle of 235 horses.

In 1964 the PA-32, an outgrowth of the PA-28, appeared. Called the Cherokee Six, it had a larger cabin and was destined to become one of the best utility aircraft ever built. One look at the Six, and

Fig. 2-1. The PA-28 was designed for simplicity of flying and maintenance.

its PA-28 heritage can be seen. Also introduced in 1964 was a stripped down PA-28 called the 140 model. It was aimed at the training market and was produced without rear seats; it was powered with a derated O-320 engine of 140 hp.

CHEROKEE 150/160

The first Cherokee, although different in construction from the PA-22 Tri-Pacer, had a cabin layout that no doubt would have made

Fig. 2-2. Over the wing and into the cockpit, through a single tight-fitting door.

Fig. 2-3. The one-piece stabilator represents simplicity and strength.

the Tri-Pacer pilot feel right at home. The trim crank was located on the ceiling, braking was accomplished with a hand lever extending from under the instrument panel, the fuel tank selector was located on the left sidewall, flaps were operated manually, and the control wheel was similar to that found on the PA-22. The power choice was 150 or 160 hp.

Changes

1963: ☐ A 37-amp alternator replaces the generator.

☐ Optional speed fairings become available.

☐ More soundproofing is added.

1965: ☐ Fiberglass cowl is introduced.

☐ Toe brakes are available as option.

☐ Cabin ventilation is improved.

1967: ☐ Last year of production.

New Prices

| **1963:** | ☐ PA-28-150 | $10,990 |
| | ☐ PA-28-160 | $11,500 |

Fig. 2-4. The Cherokee's fixed landing gear are simple and strong, with few moving parts.

Warnings

Repairing leaky fuel tanks on pre-1967 Cherokees can be a problem. The only fix is to remove them and have them resealed. This is rather expensive, and not a do-it-yourself job.

Early Cherokees were powered with Lycoming engines using $7/16$-inch-diameter exhaust valves. There is an AD (Airworthiness

Fig. 2-5. The "Hershey bar" wing.

Directive) requiring a 500-hour periodic check of these valves. Additionally, the TBO of an engine with 7/16-inch valves is only 1,200 hours. If these valves have been replaced with 1/2-inch valves, the TBO goes to 2,000 hours, and the AD no longer applies.

CHEROKEE 180/CHALLENGER/ARCHER

The next step in PA-28 evolution was the 180-hp version, PA-28-180. Introduced in 1963, the higher-powered version made the PA-28 a full four-place aircraft. The Cherokee PA-28-180 has survived competition from the Cessna Cardinal and Hawk XP, Beech Musketeer and Sundowner, and the Gulfstream Tiger. It is considered one of the most popular airplanes ever built (Figs. 2-6, 2-7).

The PA-28-180 airframe is standard PA-28. Due to the increased power and carrying ability, the 180 can be operated as a floatplane.

Changes

1965:   ☐ Fiberglass engine cowl is installed.

         ☐ Instrument panel is redesigned.

         ☐ Toe brakes become optional.

         ☐ More leg space is designed in the rear seat.

SPECIFICATIONS
PA-28-150
Cherokee 150
1962-1967

SPEED

Top Speed at Sea Level:	139 mph
Cruise:	130 mph
Stall (w/flaps):	54 mph

TRANSITIONS

Takeoff over 50' obstacle:	1,750 ft
Ground run:	800 ft
Landing over 50' obstacle:	1,890 ft
Ground roll:	535 ft

RATE OF CLIMB AT SEA LEVEL:	660 fpm
SERVICE CEILING:	14,300 ft

FUEL CAPACITY

Standard:	36 gal
ENGINE:	Lycoming O-320-E2A
TBO:	2,000 hrs
Power:	150 hp

DIMENSIONS

Wingspan:	30 ft 0 in
Length:	23 ft 3 in
Height:	7 ft 3 in

WEIGHTS

Gross Weight:	2,150 lbs
Empty Weight:	1,205 lbs
Useful Load:	945 lbs
Baggage Allowance:	125 lbs

SPECIFICATIONS

PA-28-160

Cherokee 160

1962-1967

SPEED

Top Speed at Sea Level:	141 mph
Cruise:	132 mph
Stall: (w/flaps)	55 mph

TRANSITIONS

Takeoff over 50' obstacle:	1,700 ft
Ground Run:	775 ft
Landing over 50' obstacle:	1,890 ft
Ground Roll:	550 ft

RATE OF CLIMB AT SEA LEVEL:	700 fpm
SERVICE CEILING:	15,000 ft

FUEL CAPACITY

Standard:	36 gal
ENGINE:	Lycoming O-320-B2B
TBO:	2,000 hrs
Power:	160 hp

DIMENSIONS

Wingspan:	30 ft 0 in
Length:	23 ft 3 in
Height:	7 ft 3 in

WEIGHTS

Gross Weight:	2,200 lbs
Empty Weight:	1,210 lbs
Useful Load:	990 lbs
Baggage Allowance:	125 lbs

Fig. 2-6. The PA-28-180 is one of the most popular four-place planes ever built. This particular example is a 1972 model.

Fig. 2-7. The Archer is a later version of the PA-28-180.

1967:   ☐ Larger exhaust valves are installed in the engine, giving a TBO of 2,000 hours.

1968:   ☐ Third window is added to each side.

              ☐ Instrument panel is redesigned to ''T'' system.

              ☐ Trim wheel replaces the ceiling crank.

              ☐ Throttle quadrant is installed.

1970:   ☐ Overhead air vents are installed.

1971:	☐ Rear bench seat is replaced with individual bucket seats.
1972:	☐ Air conditioning becomes optional.
1973:	☐ The fuselage is stretched eight inches, and the wings two feet.
	☐ Name is changed to Challenger.
1974:	☐ Model is renamed the Archer.
1975:	☐ Last year of PA-28-180 production.

New Prices

1963:	☐ PA-28-180	$12,900
1973:	☐ PA-28 Challenger	$16,990
1974:	☐ PA-28 Archer	$17,990

Warnings

Cherokee 180s built prior to 1970 have a redline speed between 2,150 and 2,350 rpm. Operation in this area causes torsional vibration transmitted from the crankshaft to the propeller. This can lead to propeller blade failure in extreme cases. This problem is not unique to this particular airplane, as many other makes/models have redline rpm speeds. (For further information about propellers, see Chapter 14.)

As with the Cherokees powered with the Lycoming O-320 engines, the 180s originally suffered from engines with $\frac{7}{16}$ inch diameter exhaust valves. The same AD applies, and the same change to larger valves (with extended TBO) applies.

SPECIFICATIONS
PA-28-180
Cherokee 180
1963-1972

SPEED

Top Speed at Sea Level:	148 mph
Cruise:	142 mph
Stall (w/flaps):	57 mph

RANGE

75% power at 7,000 ft:	705 mi

TRANSITIONS

Takeoff over 50' obstacle:	1,620 ft
Ground Run:	775 ft
Landing over 50' obstacle:	1,150 ft
Ground Roll:	600 ft

RATE OF CLIMB AT SEA LEVEL: 720 fpm

SERVICE CEILING: 15,700 ft

FUEL CAPACITY

Standard:	50 gal

ENGINE: Lycoming O-360-A3A

TBO:	2,000 hrs
Power:	180 hp

DIMENSIONS

Wingspan:	30 ft 0 in
Wing Area:	160 sq ft
Length:	23 ft 4 in
Height:	7 ft 4 in
Wing Loading:	15 lb/sq ft
Power Loading:	13.3 lb/hp

WEIGHTS

Gross Weight:	2,400 lbs
Empty Weight:	1,225 lbs
Useful Load:	1,175 lbs
Baggage Allowance:	125 lbs

SPECIFICATIONS
PA-28-180
Challenger/Archer
1973-1975

SPEED

Top Speed at Sea Level:	148 mph
Cruise:	142 mph
Stall (w/flaps):	61 mph

RANGE

75% power at 7,000 ft:	690 mi

TRANSITIONS

Takeoff over 50' obstacle:	1,560 ft
Ground Run:	720 ft
Landing over 50' obstacle:	1,200 ft
Ground Roll:	635 ft

RATE OF CLIMB AT SEA LEVEL:	725 fpm
SERVICE CEILING:	14,200 ft

FUEL CAPACITY

Standard:	50 gal
ENGINE:	Lycoming O-360-A4A
TBO:	2,000 hrs
Power:	180 hp

DIMENSIONS

Wingspan:	32 ft 0 in
Wing Area:	170 sq ft
Length:	24 ft 0 in
Height:	7 ft 9 in
Wing Loading:	14.4 lb/sq ft
Power Loading:	13.6 lb/hp

WEIGHTS

Gross Weight:	2,450 lbs
Empty Weight:	1,386 lbs
Useful Load:	1,064 lbs
Baggage Allowance:	200 lbs

Production Figures and Serial Numbers
Models PA-28-150/160/180

The following list shows the production years and total number of PA-28-150/160/180 airplanes that were built, along with their serial numbers.

Production Year	Total Aircraft Produced	Serial Numbers Beginning	Ending
1962	670	28-1	28-670
1963	857	28-671	28-1528
1964	131	28-1529	28-1760
1965	1,263	28-1761	28-3024
1966	810	28-3025	28-3835
1967	541	28-3836	28-4377
1968	901	28-4378	28-5279
1969	320	28-5280	28-5600
1970	259	28-5601	28-5859
1971	234	7105001	7105234
1972	318	7205001	7205318
1973	601	7305001	7305601
1974	279	7405001	7405279
1975	259	7505001	7505259

Fig. 2-8. Piper introduced the PA-28-140 as a two-seat trainer. Most such models have had the rear seat added, converting it into the four-place plane it was designed to be.

CHEROKEE 140/CRUISER/FLITELINER

In 1964, Piper Aircraft saw the need for a two-place trainer. Already seeing the success of the PA-28, it was decided to introduce a trainer version. The new model was called the PA-28-140 (Fig. 2-8). It was a standard model-150/160 airframe with the rear seats removed and a derated O-320 Lycoming engine of 140 hp. Derating was accomplished by making a propeller pitch change and lowering the engine redline to 2,450 rpm. The 140-hp engine was dropped in 1965 and replaced by the 150-hp engine.

The 140 was much roomier in all directions than its competition, the Cessna 150. However, in handling characteristics, it is not as quick or light as the Cessna 150. It is, after all, still a four-place airframe.

Many older 140s have had rear seats added to serve as family airplanes. This is fine, as long as the pilot remembers that these planes are somewhat underpowered and pays attention to loading and density altitude problems. All 140- to 160-hp four-place airplanes are sensitive in this area.

Later changes included the factory addition of a rear seat and a name change to the Cherokee Cruiser (Fig. 2-9).

Fig. 2-9. The Cruiser model is an outgrowth of the two-seat version of the PA-28.

23

Changes

1965: ☐ Horsepower is increased from 140 to 150.

1966: ☐ Rear seats become available as an option.

1969: ☐ The Cruiser model is brought out with standard rear seats, wheel fairings, and optional avionics packages.

☐ A throttle quadrant replaces the push/pull throttle.

☐ Instrument panel is rearranged to "T" format.

☐ Circuit breakers replace fuses.

☐ Rocker switches replace toggle switches.

1970: ☐ Dynafocal engine mount is introduced for noise/vibration reduction.

☐ Overhead air vents are installed.

☐ Six-way adjustable front seats are installed.

1971: ☐ Headrests and inertial-reel harnesses are added.

☐ The two-seat-only Fliteliner is introduced as a stripped down PA-28-140 for trainer use in the Piper Flite Centers.

1972: ☐ Air conditioning is offered as an option.

1973: ☐ Toe brakes are installed.

1977: ☐ Last year of production.

New Prices

1964:	☐ PA-28-140	$8,500
1969:	☐ PA-28-140 Cruiser	$9,600
1971:	☐ PA-28-140 Cruiser	$10,990
1977:	☐ PA-28-140 Cruiser	$17,170

Identification

The early Cherokees can be distinguished from the Cherokee 140s by their swing-up baggage door on the right side of the fuselage. The 140 models do not have this.

Warnings

The Cherokee 140s are not full four-place airplanes. If you carry a full load of passengers, you will not carry a full load of fuel. This is a very important item to consider in hot weather.

SPECIFICATIONS
PA-28-140
Cherokee 140/Cruiser
1964-1977

SPEED

Top Speed at Sea Level:	142 mph
Cruise:	135 mph
Stall (w/flaps):	55 mph

RANGE

75% power at 7,000 ft:	720 mi

TRANSITIONS

Takeoff over 50' obstacle:	1,700 ft
Ground Run:	800 ft
Landing over 50' obstacle:	1,090 ft
Ground Roll:	535 ft
RATE OF CLIMB AT SEA LEVEL:	631 fpm
SERVICE CEILING:	10,950 ft

FUEL CAPACITY

Standard:	36 gal
Optional:	50 gal

PROPELLER:	Sensenich M74DM 74 in
ENGINE:	Lycoming O-320-E2A
TBO:	2,000 hrs
Power:	150 hp

DIMENSIONS

Wingspan:	30 ft 0 in
Wing Area:	160 sq ft
Length:	23 ft 4 in
Height:	7 ft 4 in
Wing Loading:	13.4 lb/sq ft

WEIGHTS

Gross Weight Early 140-hp models:	1,950 lbs
Other models:	2,150 lbs
Empty Weight:	1,274 lbs
Useful Load Early 140-hp models:	676 lbs
Other models:	876 lbs
Baggage Allowance:	200 lbs

SPECIFICATIONS
PA-28-140
Fliteliner
1971-1977

SPEED

Top Speed at Sea Level:	139 mph
Cruise:	132 mph
Stall (w/flaps):	55 mph

RANGE

75% power at 7,000 ft:	705 mi

TRANSITIONS

Takeoff over 50' obstacle:	1,700 ft
Ground Run:	800 ft
Landing over 50' obstacle:	1,090 ft
Ground Roll:	535 ft

RATE OF CLIMB AT SEA LEVEL:	630 fpm
SERVICE CEILING:	10,950 ft

FUEL CAPACITY

Standard:	36 gal
Optional:	50 gal

PROPELLER:	Sensenich M74DM 74 in
ENGINE:	Lycoming O-320-E2A
TBO:	2,000 hrs
Power:	150 hp

DIMENSIONS

Wingspan:	30 ft 0 in
Wing Area:	160 sq ft
Length:	23 ft 4 in
Height:	7 ft 4 in
Wing Loading:	13.4 lb/sq ft

WEIGHTS

Gross Weight:	2,150 lbs
Empty Weight:	1,305 lbs
Useful Load:	845 lbs
Baggage Allowance:	200 lbs

Production Figures and Serial Numbers
Model PA-28-140

The following list shows the production years and total number of PA-28-140 airplanes that were built, along with serial numbers.

Production Year	Total Aircraft Produced	Serial Numbers Beginning	Ending
1964	655	28-20001	28-20655
1965	730	28-20656	28-21386
1966	1,192	28-21387	28-22579
1967	1,532	28-22580	28-24112
1968	886	28-24113	28-24999
1969	1,401	28-25000	28-26400
1970	546	28-26401	28-26946
1971	641	7125001	7125641
1972	602	7225001	7225602
1973	674	7325001	7325674
1974	444	7425001	7425444
1975	340	7525001	7525340
1976	275	7625001	7625275
1977	290	7725001	7725290

CHEROKEE 235/CHARGER/PATHFINDER

The last step in the evolution of the original PA-28 airplanes was the mating of 235 horsepower to the standard PA-28 airframe. The result was the PA-28-235, introduced in 1964 (Fig. 2-10). Although 217 units were built during 1963, they are considered to be 1964 models. The airplane is a mover, capable of better than 160-mph cruise speed. A constant-speed propeller was optional, and a float version was available. The 235 is noted for its quiet operation, due to the installation of more sound-insulation pads and doubled windows.

Fig. 2-10. The PA-28-235 is a true four-place airplane, capable of carrying almost any load into the air.

The model 235 is a fine performer, offering excellent cruise speed and performance, yet without a complex retractable landing gear system. The short-field capabilities are startling.

Changes

1968: ☐ Additional soundproofing is installed.

☐ Instrument panel is redesigned.

1973: ☐ Model is renamed the Charger.

☐ Fuselage is extended 5 inches.

☐ A larger tail is installed.

☐ Gross weight is increased 100 pounds.

☐ Wingspan is increased.

1974: ☐ Model is renamed the Pathfinder.

1977: ☐ Last year of production.

New Prices

1964:	☐ PA-28-235	$15,900
1973:	☐ PA-28-235 Charger	$24,390
1974:	☐ PA-28-235 Pathfinder	$24,390
1977:	☐ PA-28-235 Pathfinder	$35,060

SPECIFICATIONS

PA-28-235

Cherokee 235

1964-1972

SPEED

Top Speed at Sea Level:	166 mph
Cruise:	156 mph
Stall:	60 mph

TRANSITIONS

Takeoff over 50' obstacle:	1,040 ft
Ground Run:	600 ft
Landing over 50' obstacle:	1,060 ft
Ground Roll:	550 ft

RATE OF CLIMB AT SEA LEVEL:	825 fpm
SERVICE CEILING:	14,500 ft

FUEL CAPACITY

Standard:	50 gal
Optional:	84 gal

ENGINE:	Lycoming O-540-B4B5
TBO:	1,800 hrs
Power:	235 hp

DIMENSIONS

Wingspan:	32 ft 0 in
Length:	23 ft 8 in
Height:	7 ft 3 in

WEIGHTS

Gross Weight:	2,900 lbs
Empty Weight:	1,435 lbs
Useful Load:	1,465 lbs
Baggage Allowance:	200 lbs

SPECIFICATIONS

PA-28-235

Charger/Pathfinder

1973-1977

SPEED

Top Speed at Sea Level:	166 mph
Cruise:	153 mph
Stall:	61 mph

TRANSITIONS

Takeoff over 50' obstacle:	1,260 ft
Ground Run:	800 ft
Landing over 50' obstacle:	1,740 ft
Ground Roll:	1,040 ft

RATE OF CLIMB AT SEA LEVEL:	800 fpm
SERVICE CEILING:	13,550 ft

FUEL CAPACITY

Standard:	82 gal
PROPELLER:	Hartzell C/S 80 in
ENGINE:	Lycoming O-540-B4B5
TBO:	1,800 hrs
Power:	235 hp

DIMENSIONS

Wingspan:	32 ft 0 in
Wing Area:	170 sq ft
Length:	24 ft 1 in
Height:	7 ft 6 in

WEIGHTS

Gross Weight:	3,000 lbs
Empty Weight:	1,592 lbs
Useful Load:	1,408 lbs
Baggage Allowance:	200 lbs

Production Figures and Serial Numbers
Model PA-28-235

The following list shows the production years and total number of PA-28-235 airplanes that were built, along with serial numbers.

Production Year	Total Aircraft Produced	Serial Numbers Beginning	Ending
1964	585	28-10000	28-10584
1965	135	28-10585	28-10719
1966	120	28-10720	28-10840
1967	159	28-10841	28-10999
1968	227	28-11000	28-11226
1969	74	28-11227	28-11300
1970	78	28-11301	28-11378
1971	28	28-7110001	28-7110028
1972	23	28-7210001	28-7210023
1973	176	28-7310001	28-7310176
1974	109	28-7410001	28-7410109
1975	135	28-7510001	28-7510135
1976	181	28-7610001	28-7610181
1977	89	28-7710001	28-7710089

OBSERVATIONS

Over the years I have flown many different PA-28 airplanes. As a result, I have formed the following opinions and impressions.

The 150-hp versions are nice planes to fly locally, and even on short cross-country flights. I think they are too slow for flights of over 1,000 miles, although I have made many such trips. In addition to being slow, due to the smaller engine, they do not have the load-carrying capabilities of the larger-engined models. They are, however, very economical to operate and maintain.

The 180-models represent what I think are the best middle-of-the-road airplanes for price and performance. They are full four-place airplanes and can also lift plenty of baggage. The slightly better cruise will get you to your destination quicker than the 150-hp versions. They fly the same as the 150/160-models and are nearly as economical to operate (slightly more fuel burn)—and just as economical to maintain.

I really enjoy the old 235s; they are real muscle planes. The 235 is capable of lifting anything you can put in it. The drawback, however, is a large thirst for avgas. For the pilot wanting hot-rod performance, yet simplicity of operation (no retractable landing gear), the PA-28 235 model is recommended. The later models, called Charger and Pathfinder, offer more leg room in the back seat. Of note, the 235 models don't stall with power on; they keep right on mushing.

All of these planes have manual flaps, a feature I much prefer over the more modern electric systems. Flap control and movement is instant, not delayed by a motor-driven system. Also, the simplicity of manual operation means less expense for maintenance. The flaps, however, are not what the typical Cessna pilot is accustomed to. The flap area is not as large as that of the 172/182-series and does not have the Cessna's dramatic effect. However, they are adequate.

Electric trim has been installed on many of these airplanes. It is actuated from a small switch on the control wheel. I found it to be a real handy device for trimming to level flight and setting up landing approaches.

The landing gear has a footprint of 10 feet, making the plane very easy to control during crosswind landings. However, there is one glitch: The nosewheel is connected directly to the rudder pedals. This means if the nosewheel makes contact with the runway while you still have some rudder cranked in, the plane will swerve. In some instances, this could set you up for a ground loop.

Brakes vary from one PA-28 to another. The early models all were built with a lever-operated brake. This under-panel lever applied equal braking pressure to both main wheels. There was no differential braking. The later models had toe brakes, yet retained the lever brake. Many of the early airplanes have been modified with the addition of toe brakes. I personally don't feel the brakes on the PA-28s are as effective as those found on the Cessnas.

The engine cover is easily removed for engine inspection, a nice feature when you tie down outside. You know the outside—where airplane owners thoughtfully house birds in the shade of the cowling.

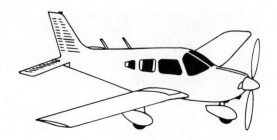

Chapter 3

The Improved Indians

In 1974, the largest single change in the PA-28 series of Piper aircraft was made. This was the introduction of an improved wing, normally referred to as the "Warrior wing."

The Warrior wing was 35 feet long and featured tapered outer panels. The new wing replaced the older constant-chord wing, often called the "Hershey bar" wing. The improved wing had a higher aspect ratio than the older version, thereby lifting more load for the same square footage of wing surface (Fig. 3-1).

The tapered wing design also changed the stall characteristics of the PA-28s. The stalls became more controllable and gentle, due to the ability of the outboard wing sections to retain lift at speeds below the point where the inboard sections stall.

The new tapered wing was added to all PA-28-type models over a period of several years. Name and model changes were made to reflect these changes.

WARRIOR

In 1974 the Warrior airplane was introduced. It was a meld of the PA-28-180 and the new, improved, tapered wing (Fig. 3-2). The object of the new design was to compete with Cessna's ever-popular Model 172 Skyhawk series.

The new PA-28-151 was able to beat the competition in useful load by a few pounds, but the big factor was the increased cabin

Fig. 3-1. The Warrior wing, longer and with a modified shape, is found on all late-model PA-28s.

room. The shoulder room was over four inches greater than the Cessna—a real plus.

Changes

1976: ☐ Last year of production of the 150-hp Warrior.

1977: ☐ The Warrior II is introduced with a 160-horse-power engine.

1978: ☐ Aerodynamic cleanup is accomplished, including landing-gear strut fairings and streamlined cuffs between the wing and main wheels. These modifications result in a 7-knot increase in cruise speed.

1979: ☐ New propeller spinner is introduced.

1980: ☐ "Expensive style" velour interiors become optional.

New Prices

1974:	☐ PA-28-151	$14,990
1978:	☐ PA-28-161	$20,820
1985:	☐ PA-28-161	$46,240
1987:	☐ PA-28-161	$69,860

Fig. 3-2. The Warrior II with its 160-hp engine makes a fine family plane. Warriors are very affordable, and won't take you to the poorhouse at maintenance time.

35

SPECIFICATIONS
PA-28-151
Warrior
1974-1977

SPEED

Top Speed at Sea Level:	135 mph
Cruise:	133 mph
Stall (landing configuration):	58 mph

RANGE

75% power at 7,000 ft:	720 mi

TRANSITIONS

Takeoff over 50' obstacle:	1,760 ft
Ground Run:	1,065 ft
Landing over 50' obstacle:	1,115 ft
Ground Roll:	595 ft

RATE OF CLIMB AT SEA LEVEL:	649 fpm
SERVICE CEILING:	12,700 ft

FUEL CAPACITY

Standard:	50 gal
ENGINE:	Lycoming O-320-E3D
TBO:	2,000 hrs
Power:	150 hp

DIMENSIONS

Wingspan:	35 ft 0 in
Wing Area:	170 sq ft
Length:	23 ft 9 in
Height:	7 ft 4 in
Wing Loading:	13.67 lb/sq ft
Power Loading:	15.5 lb/hp

WEIGHTS

Gross Weight:	2,325 lbs
Empty Weight:	1,301 lbs
Useful Load:	1,024 lbs
Baggage Allowance:	200 lbs

SPECIFICATIONS
PA-28-161
Warrior II
1977-Present

SPEED

Top Speed at Sea Level:	146 mph
Cruise:	141 mph
Stall (landing configuration):	51 mph

RANGE

75% power at 7,000 ft:	679 mi

TRANSITIONS

Takeoff over 50' obstacle:	1,650 ft
Ground Run:	1,050 ft
Landing over 50' obstacle:	1,160 ft
Ground Roll:	625 ft

RATE OF CLIMB AT SEA LEVEL:	644 fpm
SERVICE CEILING:	11,000 ft

FUEL CAPACITY

Standard:	50 gal

ENGINE:	Lycoming O-320-D3G
TBO:	2,000 hrs
Power:	160 hp

DIMENSIONS

Wingspan:	35 ft 0 in
Wing Area:	170 sq ft
Length:	23 ft 9 in
Height:	7 ft 4 in
Wing Loading:	14.4 lb/sq ft
Power Loading:	15.3 lb/hp

WEIGHTS

Gross Weight:	2,325 lbs
Empty Weight:	1,391 lbs
Useful Load:	934 lbs
Baggage Allowance:	200 lbs

Production Figures and Serial Numbers
Models PA-28-151/161

The following list shows the production years and total number of PA-28-151/161 airplanes that were built, along with serial numbers.

Production Year	Total Aircraft Produced	Serial Numbers Beginning	Ending
		Model PA-28-151	
1974	703	28-7415001	28-7415703
1975	449	28-7515001	28-7515449
1976	435	28-7615001	28-7615435
1977	314	28-7715001	28-7715314
		Model PA-28-161	
1977	323	28-7716001	28-7716323
1978	680	28-7816001	28-7816680
1979	598	28-7916001	28-7916598
1980	373	28-8016001	28-8016373
1981	322	28-8116001	28-8116322
1982	226	28-8216001	28-8216226
1983	109	28-8316001	28-8316109
1984	131	28-8416001	28-8416131
1985	99	28-8516001	28-8516099
1986	10	28-8616001	2816010
1987	n/a	2816011	n/a

ARCHER II

The Archer II was introduced in 1976. Like the Warrior, the Archer II is a melding of the tapered wing and the 180 airframe (Fig. 3-3). It is to the Warrior what the Model 180 was to the Cherokee

Fig. 3-3. The Archer II is very popular. With its 180-hp engine, it offers improved performance over the Warrior. (courtesy Piper Aircraft)

150/160, an improved model only in the horsepower department. However, this increase in horsepower from 150/160 to 180 is desirable to many pilots.

The Model 181, Archer II, like its predecessor the 180, is a full four-passenger airplane. This equates to four people, their baggage, and a full load of fuel on board. For the lightly loaded plane, the 180 horses provide much improved cruising speed and performance over the 160-hp plane.

Changes

1978:   ☐ Aerodynamic cleanup is accomplished, including landing-gear strut fairings and streamlined cuffs between the wing and main wheels. These modifications result in a 7-knot increase in cruise speed.

1980:   ☐ "Expensive style" velour interiors become optional.

New Prices

1976:	☐ PA-28-181	$23,980
1980:	☐ PA-28-181	$34,010
1985:	☐ PA-28-181	$55,320
1987:	☐ PA-18-181	$75,450

SPECIFICATIONS
PA-28-181
Archer II
1976-Present

SPEED

Top Speed at Sea Level:	148 mph
Cruise:	145 mph
Stall (landing configuration):	54 mph

RANGE

75% power with 45 min. reserve:	690 mi

TRANSITIONS

Takeoff over 50' obstacle:	1,660 ft
Ground Run:	870 ft
Landing over 50' obstacle:	1,390 ft
Ground Roll:	925 ft

RATE OF CLIMB AT SEA LEVEL:	735 fpm
SERVICE CEILING:	13,650 ft

FUEL CAPACITY

Standard:	50 gal
ENGINE:	Lycoming O-360-A4M
TBO:	2,000 hrs
Power:	180 hp

DIMENSIONS

Wingspan:	35 ft 0 in
Wing Area:	170 Sq ft
Length:	23 ft 9 in
Height:	7 ft 4 in
Wing Loading:	15 lb/sq ft
Power Loading:	14.2 lb/hp

WEIGHTS

Gross Weight:	2,550 lbs
Empty Weight:	1,413 lbs
Useful Load:	1,137 lbs
Baggage Allowance:	200 lbs

Production Figures and Serial Numbers
Model PA-28-181

The following list shows the production years and total number of PA-28-181 airplanes that were built, along with serial numbers.

Production Year	Total Aircraft Produced	Serial Numbers Beginning	Ending
1976	476	28-7690001	28-7690476
1977	607	28-7790001	28-7790607
1978	551	28-7890001	28-7890551
1979	589	28-7990001	28-7990589
1980	372	28-8090001	28-8090372
1981	313	28-8190001	28-8190313
1982	178	28-8290001	28-8290178
1983	90	28-8390001	28-8390090
1984	114	28-8490001	28-8490114
1985	92	28-8590001	28-8590092
1986	21	28-8690001	2890021
1987	n/a	2890022	n/a

DAKOTA/TURBO DAKOTA

Introduced in 1979, the Dakota is the highest-powered current PA-28. It is basically a taper-wing version of the PA-28-235 (Fig. 3-4).

The Dakota is 11 inches longer than the other two late-model PA-28s, due to the longer six-cylinder engine, and offers the best performance of all the fixed-gear PA-28s. It is also the most expensive to purchase and operate. This is due to the 235-hp engine and the constant-speed propeller. Both cause increases in maintenance, and the 235 horses must be fed. To give an idea of

Fig. 3-4. The 235-hp Dakota is capable of carrying heavy loads, just like the older PA-28 235s.

the increase in cost, Piper, in 1982, released calculated operating costs for all three PA-28 taper wings:

Warrior II	$32.45/hr
Archer II	$33.13/hr
Dakota	$44.07/hr

In all fairness to the Dakota, you must remember that you can go farther in an hour at 144 knots than you can at 123 knots.

The Dakota also was offered in a turbocharged version, giving better high-altitude performance. Called the Turbo Dakota and carrying the model number PA-28-201T, it was powered by a 200-horsepower turbocharged Continental engine. The Turbo Dakota was only offered in 1979.

Changes

1979: ☐ Turbo Dakota is offered for this year only.

1980: ☐ "Expensive style" velour interiors become optional.

New Prices

1979:	☐ PA-28-236	$39,910
1979:	☐ PA-28-201T	$41,980
1985:	☐ PA-28-236	$72,770
1987:	☐ PA-28-236	$106,920

SPECIFICATIONS

PA-28-236

Dakota

1979-Present

SPEED

Top Speed at Sea Level:	170 mph
Cruise:	166 mph
Stall (landing configurations):	64 mph

RANGE

75% power with 45 min. reserve:	748 mi

TRANSITIONS

Takeoff over 50' obstacle:	1,216 ft
Ground Run:	886 ft
Landing over 50' obstacle:	1,725 ft
Ground Roll:	825 ft

RATE OF CLIMB AT SEA LEVEL:	1,115 fpm
SERVICE CEILING:	17,500 ft

FUEL CAPACITY

Standard:	72 gal
ENGINE:	Lycoming O-540-J3A5D
TBO:	2,000 hrs
Power:	235 hp

DIMENSIONS

Wingspan:	35 ft 0 in
Wing Area:	170 sq ft
Length:	24 ft 8 in
Height:	7 ft 4 in

WEIGHTS

Gross Weight:	3,000 lbs
Empty Weight:	1,610 lbs
Useful Load:	1,390 lbs
Baggage Allowance:	200 lbs

SPECIFICATIONS
PA-28-201T
Turbo Dakota
1979

SPEED

Cruise (75% /12,000 ft):	166 mph
Cruise (65% /12,000 ft):	157 mph
Cruise (full/20,000 ft):	177 mph
Cruise (65% /20,000 ft):	174 mph
Stall (landing configuration):	67 mph

TRANSITIONS

Takeoff over 50' obstacle:	1,402 ft
Ground Run:	963 ft
Landing over 50' obstacle:	1,697 ft
Ground Roll:	861 ft

RATE OF CLIMB AT SEA LEVEL: 902 fpm

SERVICE CEILING: 20,000 ft

FUEL CAPACITY

Standard:	72 gal

ENGINE: Continental TSIO-360-FB

TBO:	1,400 hrs
Power:	200 hp

DIMENSIONS

Wingspan:	35 ft 0 in
Wing Area:	170 sq ft
Length:	24 ft 8 in
Height:	7 ft 4 in
Wing Loading:	17 lb/sq ft
Power Loading:	14.5 lb/hp

WEIGHTS

Gross Weight:	2,900 lbs
Empty Weight:	1,563 lbs
Useful Load:	1,337 lbs
Baggage Allowance:	200 lbs

Production Figures and Serial Numbers
Models PA-28-201T/236

The following list shows the production years and total number of Dakota airplanes that were built, along with serial numbers.

Production Year	Total Aircraft Produced	Serial Numbers Beginning	Ending
Model PA-28-201T			
1979	90	28-7921002	28-7921091
Model PA-28-236			
1979	335	28-7911001	28-7911335
1980	151	28-8011001	28-8011151
1981	96	28-8111001	28-8111096
1982	45	28-8211001	28-8211045
1983	25	28-8311001	28-8311025
1984	31	28-8411001	28-8411031
1985	20	28-8511001	28-8511020
1986	4	28-8611001	2811004
1987	n/a	2811005	n/a

OBSERVATIONS

In flying the improved Indians, those with the "Warrior" wing, I've noted the following:

All exhibited improved stall and slow-flight characteristics over the older "fat-winged" PA-28s. Stalls amounted to nothing more than a slight rocking motion, with no hard break. The ailerons remain effective at all times. A pilot really has to work to get into trouble with these planes.

The new wings seem to develop considerably more lift than the old constant-chord wings. This is particularly evident on landings. Either be slow at flare, or take the scenic float down the runway.

I was completely amazed at the ease of operation of the Warrior II in and out of a 1,000-foot grass strip. I could get the plane in the air at about 44 knots (using about 650 feet of runway). Of course, it was lightly loaded.

Ground handling feels better on the new planes. The nosewheel is connected to the rudder pedals with springs, rather than the direct connection found on the older models.

I enjoyed the improved ventilation in the cabin. On the ground it was not usually necessary to operate with the door held partially open, even on hot days. I did get to fly an air-conditioned version, but I personally feel this is something needed on the ground only. It's just another potential maintenance expense, and it reduces your useful load. I shut it off just before takeoff, and never turned it on again.

I noticed that the Warriors and Archers are available for rentals at almost every FBO. This must say something about their easy flying qualities and low maintenance costs.

The Dakota will give the slow end of the retractable market a run for its cruise money. The plane cruises out nicely, yet does not have the added complexity (or insurance premium) of the retractables. Like the older PA-28-235s, it will lift anything you can squeeze through the door, something the small retractables cannot do.

It's interesting that with each model I flew, I always exceeded the factory numbers for cruise—not by much, but always by at least two knots. No manufacturer's numerical inflation here!

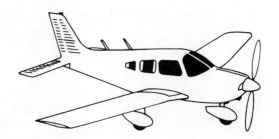

Chapter 4

The Retractables

In the mid-1950s, Piper Aircraft Corporation saw the need for a light, retractable, single-engine airplane. At that time the Beech Bonanza (production of which started in 1947) was the queen of the high-performance small planes. Piper's only four-place single was the Tri-Pacer. PA-22 tube-and-fabric was a far cry from the sleek, all-metal Bonanza.

COMANCHE

The first of the new generation Piper retractable singles was the PA-24. This all-metal, low-wing plane was called the Comanche (Fig. 4-1). The prototype first flew on May 23, 1956.

Designed for use in business travel, night flying, and instrument flying, the Comanche was miles ahead of the Piper PA-22 Tri-Pacer, the latter displaying a family history dating back to the prewar Cubs. The new Comanche had a retractable landing gear, tapered planform wings, and a laminar-flow profile. The vertical fin was swept back, and it had a stabilator.

Compared to the Bonanza, the PA-24 flew about 17 knots slower, but had better range, faster climb, weighed less, and could deliver a greater useful load. It was also about $7,000 cheaper than the Bonanza.

In all, there were seven versions of the Comanche, ranging from the original 180-hp model, first introduced in 1958, to the 400-hp model tried out around 1964 (Fig. 4-2).

Fig. 4-1. The PA-24 Comanche, Piper's first single-engine retractable. It's popular, and for average dollar value, represents a lot of airplane.

The end of the line came in 1972, when hurricane Agnes caused the Susquehanna River to devastate the city of Lock Haven and the Piper plant located there. The tooling and production equipment used in building the Comanche were destroyed.

Piper decided not to resume the Comanche line, as the cost of retooling would have been prohibitive. Additionally, PA-24 sales were down, and the Arrow, also a light retractable (production of which started in 1967), was selling well.

Fig. 4-2. Comanche Model 400. Four hundred stampeding horses up front turn this PA-24 into a four-place fighter. (courtesy Mike Keedy)

Changes

1958: ☐ 250-hp PA-24-250 enters production.

1959: ☐ Single-axis autopilot becomes standard.

1960: ☐ Reclining seats are added.

1961: ☐ Optional 15 gallon wing tip tanks become available.

1962: ☐ Redesigned electric flaps are introduced.

1963: ☐ Last year of PA-24-180 and -250 production.

1964: ☐ Start of Model 260 production.

☐ Start of Model 400 production.

☐ Electric stabilator trim becomes optional.

☐ Single-fork main landing gear are introduced.

1966: ☐ The 260B is introduced with a 6-inch extension of the fuselage, a third set of side windows, and optional fifth and sixth seats.

☐ Improved noise insulation is installed.

1969: ☐ The 260C, with the "tiger shark" cowling, enters production.

1970: ☐ The Turbo 260C appears, with a manually controlled Rayjay turbocharger.

1972: ☐ Comanche production ends.

New Prices

1958:	☐ PA-24-180	$14,500
1958:	☐ PA-24-250	$17,900
1964:	☐ PA-24-400	$28,750
1965:	☐ PA-24-260	$22,600
1966:	☐ PA-24-260B	$24,680
1969:	☐ PA-24-260C	$27,500
1970:	☐ PA-24-260C Turbo	$33,740

Identification

- [ ] In 1961 a small air scoop was installed on the top of the fuselage.

- [ ] 1964 saw the air scoop replaced with a duct in the dorsal fin. In the same year, double-paned side windows were installed.

- [ ] A third set of side windows was added in 1966.

- [ ] In 1969 the streamlined engine cowl was added.

- [ ] All 400s have leather interiors.

Warnings

The Comanches have been the object of many ADs, some very expensive to comply with:

72-22-5   Requires the installation of balancing weights to the tail control surfaces, or placarding the V_{ne} (never exceed speed) to 163 knots, and the V_{no} (maximum structural cruise speed) to 145 knots.

75-12-6   Mandates a 100-hour inspection of the vertical-fin's forward-spar fuselage attach point for cracks.

82-19-1   Calls for a 100-hour inspection of the wing's lower-main-spar caps and the upper-main-spar attachment.

77-13-21   Requires an inspection of the landing gear system every 1,000 hours, and a replacement of the bungee cords every 500 hours/3 years.

For more AD information on the PA-24, see Chapter 8.

SPECIFICATIONS

PA-24-180

Comanche 180

1958-1963

SPEED

Top Speed at Sea Level:	167 mph
Cruise:	160 mph
Stall (w/flaps):	61 mph

RANGE

75% power at 8,000 ft, no reserve:	900 mi
fuel consumption:	10.5 gph

TRANSITIONS

Takeoff over 50′ obstacle:	2,240 ft
Ground Run:	750 ft
Landing over 50′ obstacle:	1,025 ft
Ground Roll:	600 ft

RATE OF CLIMB AT SEA LEVEL: 910 fpm

SERVICE CEILING: 18,500 ft

FUEL CAPACITY

Standard:	60 gal
Optional:	90 gal

ENGINE: Lycoming O-360-A1A

TBO:	2,000 hrs
Power:	180 hp

DIMENSIONS

Wingspan:	36 ft 0 in
Length:	24 ft 9 in
Height:	7 ft 3 in

WEIGHTS

Gross Weight:	2,550 lbs
Empty Weight:	1,475 lbs
Useful Load:	1,075 lbs

SPECIFICATIONS

PA-24-250

Comanche 250

1958-1963

SPEED

Top Speed at Sea Level:	190 mph
Cruise:	181 mph
Stall (w/flaps):	61 mph

RANGE

75% power at 8,000 ft, no reserve:	780 mi
fuel consumption:	14.0 gph

TRANSITIONS

Takeoff over 50' obstacle:	1,650 ft
Ground Run:	750 ft
Landing over 50' obstacle:	1,280 ft
Ground Roll:	600 ft

RATE OF CLIMB AT SEA LEVEL:	1,350 fpm
SERVICE CEILING:	20,000 ft

FUEL CAPACITY

Standard:	60 gal
Optional:	90 gal

ENGINE:	Lycoming O-540-A1A5
TBO:	2,000 hrs
Power:	250 hp

DIMENSIONS

Wingspan:	36 ft 0 in
Length:	24 ft 9 in
Height:	7 ft 3 in

WEIGHTS

Gross Weight:	2,800 lbs
Empty Weight:	1,600 lbs
Useful Load:	1,200 lbs

SPECIFICATIONS

PA-24-260

Comanche 260

1964-1965

SPEED

Top Speed at Sea Level:	194 mph
Cruise:	186 mph
Stall (w/flaps):	61 mph

RANGE

75% power at 8,000 ft, no reserve:	730 mi
fuel consumption:	14.1 gph

TRANSITIONS

Takeoff over 50' obstacle:	1,040 ft
Ground Run:	760 ft
Landing over 50' obstacle:	1,420 ft
Ground Roll:	655 ft

RATE OF CLIMB AT SEA LEVEL:	1,500 fpm
SERVICE CEILING:	10,600 ft

FUEL CAPACITY

Standard:	60 gal
Optional:	90 gal

ENGINE:	Lycoming O-540-E4A5
TBO:	2,000 hrs
Power:	260 hp

DIMENSIONS

Wingspan:	36 ft 0 in
Length:	24 ft 9 in
Height:	7 ft 3 in

WEIGHTS

Gross Weight:	2,900 lbs
Empty Weight:	1,700 lbs
Useful Load:	1,200 lbs

SPECIFICATIONS
PA-24-260B/C
Comanche 260B & C
1966-1972

SPEED

Top Speed at Sea Level:	194 mph
Cruise:	182 mph
Stall (w/flaps):	61 mph

RANGE

75% power at 8,000 ft, no reserve:	730 mi
fuel consumption:	14.1 gph

TRANSITIONS

Takeoff over 50' obstacle:	1,260 ft
Ground Run:	760 ft
Landing over 50' obstacle:	1,435 ft
Ground Roll:	655 ft

RATE OF CLIMB AT SEA LEVEL:	1,370 fpm
SERVICE CEILING:	20,000 ft

FUEL CAPACITY

Standard:	60 gal
Optional:	90 gal

ENGINE:	Lycoming IO-540-D4A5
TBO:	2,000 hrs
Power:	260 hp

DIMENSIONS

Wingspan:	36 ft 0 in
Length:	25 ft 3 in
Height:	7 ft 3 in

WEIGHTS

Gross Weight:	3,100 lbs
Empty Weight:	1,728 lbs
Useful Load:	1,372 lbs

SPECIFICATIONS
PA-24-260TC
Comanche 260 Turbo C
1970-1972

SPEED

Top Speed at Sea Level:	195 mph
Cruise:	185 mph
Stall (w/flaps):	61 mph

RANGE

75% power at 8,000 ft, no reserve:	830 mi
fuel consumption:	15.4 gph

TRANSITIONS

Takeoff over 50' obstacle:	1,400 ft
Ground Run:	820 ft
Landing over 50' obstacle:	1,465 ft
Ground Roll:	690 ft

RATE OF CLIMB AT SEA LEVEL:	1,320 fpm
SERVICE CEILING:	25,000 ft

FUEL CAPACITY

Standard:	60 gal
Optional:	90 gal

ENGINE:	Lycoming IO-540-N1A5
TBO:	2,000 hrs
Power:	260 hp

DIMENSIONS

Wingspan:	36 ft 0 in
Length:	25 ft 8 in
Height:	7 ft 3 in

WEIGHTS

Gross Weight:	3,200 lbs
Empty Weight:	1,894 lbs
Useful Load:	1,306 lbs

SPECIFICATIONS
PA-24-400
Comanche 400
1964-1965

SPEED

Top Speed at Sea Level:	223 mph
Cruise:	213 mph
Stall (w/flaps):	68 mph

RANGE

75% power at 8,000 ft, no reserve:	1,000 mi
fuel consumption:	23.0 gph

TRANSITIONS

Takeoff over 50' obstacle:	1,500 ft
Ground Run:	980 ft
Landing over 50' obstacle:	1,820 ft
Ground Roll:	1,180 ft

RATE OF CLIMB AT SEA LEVEL: 1,600 fpm

SERVICE CEILING: 19,500 ft

FUEL CAPACITY

Standard:	100 gal
Optional:	130 gal

ENGINE: Lycoming IO-720-A1A

TBO:	1,800 hrs
Power:	400 hp

DIMENSIONS

Wingspan:	36 ft 0 in
Length:	25 ft 8 in
Height:	7 ft 3 in

WEIGHTS

Gross Weight:	3,600 lbs
Empty Weight:	2,110 lbs
Useful Load:	1,490 lbs

Production Figures and Serial Numbers

Models PA-24-180/250/260/400

The following list shows the production years and total number of PA-24-180/250/260/400 airplanes that were built, along with serial numbers.

Production Year	Total Aircraft Produced	Serial Numbers Beginning	Ending
1958	336	24-1	24-336
1959	1,140	24-337	24-1476
1960	822	24-1477	24-2298
1961	545	24-2299	24-2843
1962	441	24-2844	24-3284
1963	273	24-3285	24-3557
1964	130	24-3558	24-3687
1965	301	24-4000	24-4299
1966	359	24-4300	24-4658
1967	91	24-4661	24-4751
1968	52	24-4752	24-4803
1969	97	24-4804	24-4900
1970	50	24-4901	24-4950
1971	50	24-4951	24-5000
1972	28	24-5001	24-5028

Observations

The Comanches were great airplanes in their day; however, that day started 30 years ago. I think they are still good airplanes, but maintenance can be a problem. Some parts are no longer produced, and procurement of replacements can lead to lots of downtime.

Flying the PA-24s is a joy, and, for many proud owners, well worth the maintenance problems encountered along the way.

There are only a few true Comanche 180s left, as many have been converted to high power. Those still around are quite economical in direct operational costs, have good speed, and handle lightly, but are limited in their carrying ability. I like the 180 model, because it's cheap to fly, and engine maintenance won't eat you out of house and home.

The heavy models (260 and 400) are heavy on the controls, and require the use of trim for just about everything, unless you're a weightlifter. But that's not something unique to the PA-24s; most "heavy" planes share the need for trim.

Stalls in the 180 are quite mild—really just a gentle dropping of the nose until airspeed is recovered. This is true for both power-on and power-off stalls. As the model numbers go up, the weight goes up, and the stalls become more businesslike, but never too difficult to handle.

Due to the very low profile (short landing gear), there is a very pronounced ground effect when landing. Also, the nosewheel sticks down farther than the mains. This means you must fully flare on landings to prevent wheelbarrowing.

Approaches are not really hot, as with some other makes/models of retractables. A tame 85-95 mph does just fine on final, slowing to about 80 over the fence.

Flying the 400 gives me a real thrill, as I'm a power hog. Flying solo, it climbs like a fighter. However, there is a real drawback with the 400: Those 400 horses eat fuel at an alarming rate, often exceeding 16 gph—almost enough to keep OPEC happy.

THE PA-28R ARROWS

In 1967, Piper introduced a new light retractable, the Arrow. This plane caused the slowing of Comanche sales, and was the deciding factor against restarting PA-24 production after the 1972 flood.

Fig. 4-3. The Arrow, with automatic retractable landing gear, replaced the Comanche.

The Arrow 180, the first in the Arrow series, was basically a PA-28-180 with retractable landing gear (Fig. 4-3). With the fold-away gear, an additional 20 mph is picked up over the fixed-gear 180s.

At the time of the Arrow's introduction, the light-retractable market was occupied only by Mooney Aircraft. The last light retractable made by any of the big three (Piper, Cessna, and Beech) had been the Comanche 180. All others had been much larger in the horsepower department, hence, much heavier in weight, larger in dimension, and higher cost of purchase, operation, and maintenance.

Possibly the most significant development found on the Arrow is the idiot-proof landing-gear system. It's often said that there are only two kinds of retractable pilots: those who have landed wheels-up, and those who will. Taking this observation into account, Piper installed an automatic landing-gear actuation system on the Arrow.

"Automatic" refers to a system that measures impact air pressure from a probe on the left side of the cabin (Fig. 4-4). The

Fig. 4-4. This port is the actuator for the Arrow's landing gear.

air pressure, via the probe, controls a hydraulic valve that retains the gear in an up position. As the impact air pressure drops, caused by airspeed reduction, the hydraulic valve releases the gear, which free-falls into the down-and-locked position.

The landing gear is automatically lowered when the airplane is operating slower than 105 mph, and is retracted only when the plane passes 85 mph at full throttle.

The automatic system can be overridden, as all systems must be from time to time. A small lever between the seats allows the gear to remain retracted at any speed. In earlier models this lever was spring-actuated, and you had to keep your hand on it all the time when you wished to override the automatic system. This was somewhat inconvenient when practicing slow flight, stalls, etc. The later models will latch in the off position. When so locked, a warning light will flash on the instrument panel.

The Arrow's automatic landing gear system is so well thought of that the insurance industry will normally insure a low-time Arrow pilot with little or no retractable time.

Although the Arrow does not compete with the heavier retractables in the speed department, it is not as thirsty for fuel, either. Generally, Arrow pilots can expect 160-165 mph, and operate on 10-12 gallons of fuel per hour.

The most recent version of the Arrow is the Turbo Arrow IV (Fig. 4-5).

Fig. 4-5. The Turbo Arrow IV is the latest in the PA-28R series. (courtesy Piper Aircraft)

Changes

1969: ☐ A 200-horsepower engine becomes optional.

1971: ☐ Last year for production of the Arrow 180.

1972: ☐ The Arrow II is introduced, with a five-inch longer fuselage and a wider door.

☐ Wingspan is increased two feet.

☐ Baggage capacity is increased by 50 pounds.

☐ A locking landing-gear override lever is installed.

1977: ☐ The Arrow III, with the Warrior tapered wing, begins production.

☐ Newly introduced is the turbocharged Turbo Arrow III.

1979: ☐ The fourth version of the Arrow, the T-tailed Arrow IV, starts production.

☐ The Turbo Arrow IV is also available.

New Prices

1967:	☐ PA-28R-180	Arrow	$16,900
1969:	☐ PA-28R-200	Arrow	$19,980
1972:	☐ PA-28R-200B	Arrow II	$23,500
1977:	☐ PA-28R-201	Arrow II	$37,850
1977:	☐ PA-28R-201T	Turbo Arrow III	$41,800
1979:	☐ PA-28RT-201	Arrow IV	$44,510
1979:	☐ PA-28RT-201T	Turbo Arrow IV	$49,150
1985:	☐ PA-28RT-201	Arrow IV	$88,920
1987:	☐ PA-28RT-201T	Turbo Arrow IV	$124,400

Warnings

Remember that retractable gear will increase your costs of maintenance. There is more mechanical operation, and more moving parts. These parts will wear and need service and/or replacement at some point in time (Fig. 4-6). Additionally, there is extra cost at annual inspection time for jacking the plane up to inspect and cycle the landing gear.

For AD information on the Arrows, see Chapter 8.

Fig. 4-6. These photos show the Arrow's main- and nose-gear assemblies. Fold-away gear can give a 20-mph increase in cruising speed, but they require more maintenance than fixed gear.

SPECIFICATIONS

PA-28R-180

Arrow 180

1967-1971

SPEED
Top Speed at Sea Level:	170 mph
Cruise:	162 mph
Stall (w/flaps):	61 mph

TRANSITIONS
Takeoff over 50' obstacle:	1,240 ft
Ground Run:	820 ft
Landing over 50' obstacle:	1,340 ft
Ground Roll:	776 ft

RATE OF CLIMB AT SEA LEVEL:	875 fpm
SERVICE CEILING:	15,000 ft

FUEL CAPACITY
Standard:	50 gal

ENGINE:	Lycoming IO-360-B1E
TBO:	2,000 hrs
Power:	180 hp

DIMENSIONS
Wingspan:	30 ft 0 in
Length:	24 ft 2 in
Height:	8 ft 0 in

WEIGHTS
Gross Weight:	2,500 lbs
Empty Weight:	1,380 lbs
Useful Load:	1,120 lbs

SPECIFICATIONS

PA-28R-200

Arrow 200

1969-1971

SPEED

Top Speed at Sea Level:	176 mph
Cruise:	166 mph
Stall (w/flaps):	64 mph

RANGE

75% power:	810 mi

TRANSITIONS

Takeoff over 50' obstacle:	1,600 ft
Ground Run:	770 ft
Landing over 50' obstacle:	1,380 ft
Ground Roll:	780 ft

RATE OF CLIMB AT SEA LEVEL:	910 fpm
SERVICE CEILING:	16,000 ft

FUEL CAPACITY

Standard:	50 gal
ENGINE:	Lycoming IO-360-B1E
TBO:	1,800 hrs
Power:	200 hp

DIMENSIONS

Wingspan:	30 ft 0 in
Wing Area:	160 sq ft
Length:	24 ft 2 in
Height:	8 ft 0 in
Wing Loading:	16.3 lb/sq ft

WEIGHTS

Gross Weight:	2,600 lbs
Empty Weight:	1,459 lbs
Useful Load:	1,141 lbs

SPECIFICATIONS
PA-28R-200B
Arrow II
1972-1976

SPEED
Top Speed at Sea Level:	175 mph
Cruise:	165 mph
Stall (w/flaps):	64 mph

RANGE
55% power:	900 mi

TRANSITIONS
Takeoff over 50' obstacle:	1,600 ft
Ground Run:	770 ft
Landing over 50' obstacle:	1,380 ft
Ground Roll:	780 ft

RATE OF CLIMB AT SEA LEVEL:	900 fpm
SERVICE CEILING:	15,000 ft

FUEL CAPACITY
Standard:	50 gal

ENGINE:	Lycoming IO-360-C1C
TBO:	1,800 hrs
Power:	200 hp

DIMENSIONS
Wingspan:	32 ft 2 in
Wing Area:	170 sq ft
Length:	24 ft 7 in
Height:	8 ft 0 in

WEIGHTS
Gross Weight:	2,650 lbs
Empty Weight:	1,515 lbs
Useful Load:	1,135 lbs

SPECIFICATIONS
PA-28R-201
Arrow III
1977-1978

SPEED

Top Speed at Sea Level:	171 mph
Cruise:	164 mph
Stall (w/flaps):	61 mph

TRANSITIONS

Takeoff over 50' obstacle:	1,600 ft
Ground Run:	1,025 ft
Landing over 50' obstacle:	1,525 ft
Ground Roll:	615 ft

RATE OF CLIMB AT SEA LEVEL:	831 fpm
SERVICE CEILING:	16,200 ft

FUEL CAPACITY

Standard:	72 gal
ENGINE:	Lycoming IO-360-C1C6
TBO:	1,800 hrs
Power:	200 hp

DIMENSIONS

Wingspan:	35 ft 5 in
Length:	27 ft 0 in
Height:	8 ft 4 in

WEIGHTS

Gross Weight:	2,750 lbs
Empty Weight:	1,637 lbs
Useful Load:	1,113 lbs

SPECIFICATIONS
PA-28R-201T
Turbo Arrow III
1977-1978

SPEED

Top Speed at Sea Level:	204 mph
Cruise:	198 mph
Stall (w/flaps):	65 mph

TRANSITIONS

Takeoff over 50' obstacle:	1,620 ft
Ground Run:	1,110 ft
Landing over 50' obstacle:	1,555 ft
Ground Roll:	645 ft

RATE OF CLIMB AT SEA LEVEL:	940 fpm
SERVICE CEILING:	20,000 ft

FUEL CAPACITY

Standard:	72 gal
ENGINE:	Continental TSIO-360-F
TBO:	1,400 hrs
Power:	200 hp

DIMENSIONS

Wingspan:	35 ft 5 in
Wing Area:	170 sq ft
Length:	27 ft 4 in
Height:	8 ft 4 in
Wing Loading:	17 lb/sq ft

WEIGHTS

Gross Weight:	2,900 lbs
Empty Weight:	1,692 lbs
Useful Load:	1,208 lbs

SPECIFICATIONS

PA-28RT-201

Arrow IV

1979-Present

SPEED

Top Speed at Sea Level:	171 mph
Cruise:	164 mph
Stall (w/flaps):	61 mph

TRANSITIONS

Takeoff over 50' obstacle:	1,600 ft
Ground Run:	1,025 ft
Landing over 50' obstacle:	1,525 ft
Ground Roll:	615 ft

RATE OF CLIMB AT SEA LEVEL: 831 fpm

SERVICE CEILING: 16,200 ft

FUEL CAPACITY

Standard:	72 gal

ENGINE: Lycoming IO-360-C1C6

TBO:	1,800 hrs
Power:	200 hp

DIMENSIONS

Wingspan:	35 ft 5 in
Length:	27 ft 0 in
Height:	8 ft 4 in

WEIGHTS

Gross Weight:	2,750 lbs
Empty Weight:	1,637 lbs
Useful Load:	1,113 lbs

SPECIFICATIONS
PA-28RT-201T
Turbo Arrow IV
1979-Present

SPEED

Top Speed at Sea Level:	204 mph
Cruise:	198 mph
Stall (w/flaps):	65 mph

RANGE: 860 mi

TRANSITIONS

Takeoff over 50' obstacle:	1,620 ft
Ground Run:	1,110 ft
Landing over 50' obstacle:	1,555 ft
Ground Roll:	645 ft

RATE OF CLIMB AT SEA LEVEL: 940 fpm

SERVICE CEILING: 20,000 ft

FUEL CAPACITY

Standard:	72 gal

ENGINE: Continental TSIO-360-FB

TBO:	1,400 hrs
Power:	200 hp

DIMENSIONS

Wingspan:	35 ft 5 in
Length:	27 ft 4 in
Height:	8 ft 4 in

WEIGHTS

Gross Weight:	2,900 lbs
Empty Weight:	1,692 lbs
Useful Load:	1,208 lbs

Production Figures and Serial Numbers
PA-28R Arrows

The following list shows the production years and total number of PA-28R Arrows that were built, along with serial numbers.

Production Year	Total Aircraft Produced	Serial Numbers Beginning	Ending
Model PA-28R-180			
1967	351	28R30004	28R30351
1968	736	28R30352	28R31087
1969	163	28R31088	28R31250
1970	20	28R31251	28R31270
1971	13	28R7130001	28R30013
Model PA-28R-200			
1969	600	28R35001	28R35600
1970	220	28R35601	28R35820
1971	229	28R7135001	28R7135600
1972	320	28R7235001	28R7235229
1973	466	28R7335001	28R7335466
1974	320	28R7435001	28R7435320
1975	383	28R7535001	28R7535383
1976	545	28R7635001	28R7635545

Model PA-28R-201

| 1977 | 178 | 28R7737001 | 28R7737178 |
| 1978 | 317 | 28R7837001 | 28R7837317 |

Model PA-28R-201T

| 1977 | 427 | 28R7703001 | 28R7703427 |
| 1978 | 373 | 28R7803001 | 28R7803373 |

Model PA-28RT-201

1979	267	28R7918001	28R7918267
1980	47	28R8018001	28R8018047
1981	73	28R8118001	28R8118073
1982	26	28R8218001	28R8218026

Model PA-28RT-201T

1979	310	28R7931001	28R7931310
1980	178	28R8031001	28R8031178
1981	208	28R8131001	28R8131208
1982	69	28R8231001	28R8231069
1983	49	28R8331001	28R8331049
1984	32	28R8431001	28R8431032
1985	15	28R8531001	28R8531015
1986	1	28R8631001	28R8631001
1987	n/a	28R8631002	n/a

Observations

The Arrows are common airplanes, i.e., there are lots of them around. I found examples at each of the 10 airports that I frequent. The particular model that I flew for evaluation was the Arrow 200.

This light retractable is a delight to fly. If you've been flying a PA-28, then the transition to the PA-28R is easy. Everything is pretty much in the same place on both airplanes. The Arrow is a little faster, but other than that, if you fly a PA-28 you'll do just fine. The handling characteristics are typical Cherokee.

Take-offs come quickly, as there is plenty of acceleration. If you're in a squeeze and need even quicker liftoff, apply some flaps.

Stalls are typical Cherokee—no surprises, very gentle, with full control throughout. Lightly loaded, I couldn't get a true break no matter what I tried.

The cruise I got was 165 mph at about 10.5 gph. That's pretty economical. It also makes the miles click away. Home in west Texas is only a one-day trip from the East Coast.

I found that about 95 mph on final, with some flaps, worked fine, slowing to about 80 mph (full flaps) over the fence. Keeping a little power on all the way down will do wonders to smooth out any problems at flare.

If you really have to take the fast down elevator, just pull full flaps and slow to 90 mph IAS. The drop will be nearly 1,500 feet per minute.

I liked the manual flaps. I have never cared for electric flaps on small airplanes; they don't allow instant control.

The standard Cherokee fuel selector is used on the Arrow.

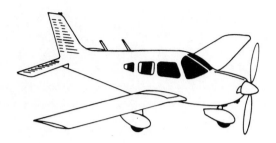

Chapter 5

The Hard-Working Sixes

In 1965, Piper Aircraft introduced yet another airplane based upon the PA-28 Cherokee. This new plane had a much longer fuselage, but otherwise closely resembled the first Cherokee. This new craft marked the start of a seven-place single-engine series.

THE CHEROKEE SIXES

The beginning model of these new "stretched-looking" airplanes was called the Cherokee Six, Model PA-32 (Fig. 5-1).

The first thing you notice when you see a PA-32 is the extreme fuselage length. The Cherokee Six is almost five feet longer than its cousin, the PA-28. It also has an additional door on the left side for the use of rear seat passengers. The airframe, although resembling a PA-28, is all original. It is *not* a modified PA-28 frame.

Originally introduced as an alternative for the very low end of the twin market, the Sixes have become very popular for charter service—not just in carrying seven passengers, but in carrying anything. Popular as air ambulances and hearses, due to the long cabin space and large doors, they are real workhorses. I have heard them referred to as "flying pickup trucks" by several FBOs who operate them in remote areas (Figs. 5-2, 5-3).

The first Sixes were powered with Lycoming 260-hp carbureted engines. Not until 1966 was the optional 300-hp fuel-injected engine offered.

Fig. 5-1. The PA-32 Cherokee Six is easily recognized by the long nose. This particular airplane is the seventh PA-32 built by Piper.

Fig. 5-2. The large rear side door of the Six makes loading large cargo easy. This door is one of the reasons Sixes are used as air ambulances.

Fig. 5-3. Storage space abounds on the PA-32s.

Changes

1966:   ☐ The 300-horsepower engine becomes available as an option.

1967:   ☐ A five-foot-wide cargo door becomes available.

1974:   ☐ A fourth side window and a rear utility door are added.

1979:   ☐ Last year of production.

New Prices

1965:	☐ PA-32-260	$18,500
1966:	☐ PA-32-300	$21,500
1974:	☐ PA-32-300	$32,290
1979:	☐ PA-32-300	$52,030

SPECIFICATIONS
PA-32-260
Cherokee Six
1965-1978

SPEED

Top Speed at Sea Level:	168 mph
Cruise:	160 mph
Stall (landing configuration):	63 mph

TRANSITIONS

Takeoff over 50' obstacle:	1,360 ft
Ground Run:	810 ft
Landing over 50' obstacle:	1,000 ft
Ground Roll:	630 ft

RATE OF CLIMB AT SEA LEVEL: 850 fpm

SERVICE CEILING: 14,500 ft

FUEL CAPACITY

Standard:	50 gal
Optional:	84 gal

ENGINE: Lycoming O-540-E4B5

TBO:	2,000 hrs
Power:	260 hp

DIMENSIONS

Wingspan:	32 ft 9 in
Length:	27 ft 9 in
Height:	7 ft 11 in

WEIGHTS

Gross Weight:	3,400 lbs
Empty Weight:	1,699 lbs
Useful Load:	1,701 lbs

SPECIFICATIONS
PA-32-300
Cherokee Six
1966-1979

SPEED

Top Speed at Sea Level:	174 mph
Cruise:	168 mph
Stall (landing configuration):	63 mph

TRANSITIONS

Takeoff over 50' obstacle:	1,500 ft
Ground Run:	1,050 ft
Landing over 50' obstacle:	1,000 ft
Ground Roll:	630 ft

RATE OF CLIMB AT SEA LEVEL:	1,050 fpm
SERVICE CEILING:	16,250 ft

FUEL CAPACITY

Standard:	50 gal
Optional:	84 gal

ENGINE:	Lycoming IO-540-K1A5
TBO:	2,000 hrs
Power:	300 hp

DIMENSIONS

Wingspan:	32 ft 9 in
Wing Area:	174.5 sq ft
Length:	27 ft 9 in
Height:	7 ft 10 in
Wing Loading:	19.5 lb/sq ft
Power Loading:	11.3 lb/hp

WEIGHTS

Gross Weight:	3,400 lbs
Empty Weight:	1,789 lbs
Useful Load:	1,611 lbs
Baggage Allowance:	200 lbs

Production Figures and Serial Numbers
Model PA-32-260/300

The following list shows the production years and total number of PA-32-260/300 airplanes that were built, along with serial numbers.

Production Year	Total Aircraft Produced	Serial Numbers Beginning	Ending
		Model PA-32-260	
1965	317	32-1	32-317
1966	534	32-319	32-852
1967	157	32-855	32-1011
1968	99	32-1012	32-1110
1969	140	32-1111	32-1250
1970	47	32-1251	32-1297
1971	23	32-7100001	32-7100023
1972	45	32-7200001	32-7200045
1973	65	32-7300001	32-7300065
1974	51	32-7400001	32-7400051
1975	43	32-7500001	32-7500043
1976	24	32-7600001	32-7600024
1977	23	32-7700001	32-7700023
1978	8	32-7800001	32-7800008

Model PA-32-300

Year	Qty	Begin	End
1966	135	32-40000	32-40135
1967	293	32-40137	32-40429
1968	136	32-40430	32-40565
1969	285	32-40566	32-40850
1970	168	32-40851	32-41018
1971	78	32-7140001	32-7140078
1972	137	32-7240001	32-7240137
1973	191	32-7340001	32-7340191
1974	170	32-7440001	32-7440170
1975	188	32-7540001	32-7540188
1976	130	32-7640001	32-7640130
1977	113	32-7740001	32-7740113
1978	202	32-7840001	32-7840202
1979	290	32-7946001	32-7940290

THE LANCES

In 1976 the Lance version of the PA-32 was introduced. The Lances were Cherokee Sixes with retractable landing gear. Until late in the 1978 model year, there were few other differences in the basic structures. The "Hershey bar" wing was utilized on all Lance models.

During the 1978 model year, a new T-tail was placed on all Lances (Fig. 5-4). The new tails caused considerable pilot rumblings and a resultant lack of favor for the later Lance II versions.

The controversy centers around performance. The performance figures for the Lance and Lance II don't depict a great difference in performance; however, most pilots disagree with the published figures. The standard complaint is that the older conventional-tailed Lances use less runway for takeoff than the T-tailed Lance IIs. Several pilots report that the Lance II needs more than 1,600 feet just to break ground. Additionally, some pilots grouse that the new tail causes nose pitch-up surprises during the takeoff run. The latter is really a "getting used to the airplane" problem.

Fig. 5-4. The T-tail Lance, an outgrowth of the first Sixes.

Changes

1976:   ☐ The Lance becomes available.

1978:   ☐ Turbocharging becomes optional on the Lance.

☐ The T-tailed Lance II is introduced mid-model year.

1979:   ☐ Last production year for the Lance series.

New Prices

1976:	☐ PA-28R-300	$49,990
1978:	☐ PA-28R-300	$54,640

Warnings

The following Airworthiness Directives are considered major in nature. As with all ADs, compliance is required.

76-11-09: Requires rerouting of the fuel lines to prevent chaffing.

76-15-08: Calls for modification of the nose gear to prevent collapse.

76-16-08: Requires replacement of the oil coolers to prevent oil starvation and engine failure.

79-26-04: Requires modification of the rudder structure to prevent cracks from developing in the skin.

SPECIFICATIONS
PA-32R-300
Lance
1976-1978

SPEED

Top Speed at Sea Level:	180 mph
Cruise:	176 mph
Stall (landing configuration):	60 mph

RANGE

75% power at 5,000 ft:	900 mi
fuel consumption:	18 gph
55% power at 13,000 ft:	990 mi
fuel consumption:	14 gph

TRANSITIONS

Takeoff over 50' obstacle:	1,660 ft
Ground Run:	960 ft
Landing over 50' obstacle:	1,708 ft
Ground Roll:	880 ft

RATE OF CLIMB AT SEA LEVEL:	1,000 fpm
SERVICE CEILING:	14,600 ft
FUEL CAPACITY (Standard):	98 gal
PROPELLER:	Hartzell C/S 80 in.
ENGINE:	Lycoming IO-540-K1G5D
TBO:	2,000 hrs
Power:	300 hp

DIMENSIONS

Wingspan:	32 ft 9 in
Wing Area:	174.5 sq ft
Length/Height:	27 ft 9 in/9 ft 0 in
Wing Loading:	20.6 lb/sq ft
Power Loading:	12.0 lb/hp

WEIGHTS

Gross/Empty Weight:	3,600 lbs/1,980 lbs
Useful Load:	1,620 lbs
Baggage Allowance:	200 lbs

SPECIFICATIONS
PA-32RT-300
Lance II
1978-1979

SPEED

Top Speed at Sea Level:	181 mph
Cruise:	178 mph
Stall (landing configuration):	60 mph

RANGE

75% power at 5,000 ft:	900 mi
fuel consumption:	18 gph
55% power at 13,000 ft:	990 mi
fuel consumption:	14 gph

TRANSITIONS

Takeoff over 50' obstacle:	1,690 ft
Ground Run:	960 ft
Landing over 50' obstacle:	1,710 ft
Ground Roll:	880 ft

RATE OF CLIMB AT SEA LEVEL:	1,000 fpm
SERVICE CEILING:	14,600 ft
FUEL CAPACITY (Standard):	98 gal
PROPELLER:	Hartzell C/S 80 in.
ENGINE:	Lycoming IO-540-K1G5D
TBO:	2,000 hrs
Power:	300 hp

DIMENSIONS

Wingspan:	32 ft 9 in
Wing Area:	174.5 Sq ft
Length/Height:	28 ft 4 in/9 ft 6 in
Wing Loading:	20.6 lb/sq ft
Power Loading:	12.0 lb/hp

WEIGHTS

Gross/Empty Weight:	3,600 lbs/2,003 lbs
Useful Load:	1,597 lbs
Baggage Allowance:	200 lbs

SPECIFICATIONS
PA-32RT-300T
Turbo Lance II
1978-1979

SPEED

Top Speed at Sea Level:	191 mph
Cruise:	186 mph
Stall (landing configuration):	60 mph

RANGE

75% power at 5,000 ft:	920 mi
fuel consumption:	18 gph
55% power at 13,000 ft:	1,070 mi
fuel consumption:	15 gph

TRANSITIONS

Takeoff over 50' obstacle:	1,660 ft
Ground Run:	960 ft
Landing over 50' obstacle:	1,710 ft
Ground Roll:	880 ft

RATE OF CLIMB AT SEA LEVEL: 1,000 fpm

SERVICE CEILING: 20,000 ft

FUEL CAPACITY (Standard): 98 gal

PROPELLER: Hartzell C/S 80 in

ENGINE: Lycoming TIO-540-S1AD

TBO:	1,800 hrs
Power:	300 hp

DIMENSIONS

Wingspan:	32 ft 9 in
Wing Area:	174.5 sq ft
Length/Height:	28 ft 11 in 9 ft 6 in
Wing Loading:	20.6 lb/sq ft
Power Loading:	12.0 lb/hp

WEIGHTS

Gross/Empty Weight:	3,600 lbs/2,071 lbs
Useful Load:	1,529 lbs
Baggage Allowance:	200 lbs

Production Figures and Serial Numbers
PA-32R Lances

The following list shows the production years and total number of PA-32R-300/RT-300/RT-300T airplanes that were built, along with serial numbers.

Production Year	Total Aircraft Produced	Serial Numbers Beginning	Ending
Model PA-32R-300			
1976	525	32R-7680001	32R-7680525
1977	548	32R-7780001	32R-7780548
1978	68	32R-7880001	32R-7880068
Model PA-32RT-300			
1978	285	32R-7885001	32R-7885285
1979	105	32R-7985001	32R-7985105
Model PA-32RT-300T			
1978	289	32R-7887001	32R-7887289
1979	126	32R-7987001	32R-7987126

THE SARATOGAS

The Saratoga is the most recent product in the PA-32 line (Fig. 5-5). As with the other late model Piper Indians, the largest single change introduced in the Saratoga is the tapered Warrior wing.

The PA-32-301 series is available in either a fixed-gear or retractable-gear configuration, the latter being called the PA-32R-301 Saratoga SP (Fig. 5-6).

Both Saratoga models were introduced in 1980, and are available with turbocharging. Saratogas come with a three-blade constant-speed propeller and a standard tail. The T-tail of the Lance II is gone.

Fig. 5-5. The Turbo Saratoga. Notice that the standard tail has returned to the PA-32. (courtesy Piper Aircraft)

Fig. 5-6. The retractable-geared Turbo Saratoga SP. (courtesy Piper Aircraft)

New Prices

1980:	☐ PA-32-301 Saratoga	$66,700
1980:	☐ PA-32-301T Turbo Saratoga	$74,900
1980:	☐ PA-32R-301 Saratoga SP	$80,200
1980:	☐ PA-32R-301T Turbo Saratoga SP	$88,400
1985:	☐ PA-32R-301T Turbo Saratoga SP	$142,100
1987:	☐ PA-32R-301T Turbo Saratoga SP	$179,820

SPECIFICATIONS

PA-32-301

Saratoga

1980-Present

SPEED

Top Speed at Sea Level:	175 mph
Cruise:	172 mph
Stall (landing configuration):	67 mph

TRANSITIONS

Takeoff over 50′ obstacle:	1,759 ft
Ground Run:	1,183 ft
Landing over 50′ obstacle:	1,612 ft
Ground Roll:	732 ft

RATE OF CLIMB AT SEA LEVEL:	990 fpm
SERVICE CEILING:	14,100 ft

FUEL CAPACITY

Standard:	102 gal

ENGINE:	Lycoming IO-540-K1G5
TBO:	2,000 hrs
Power:	300 hp

DIMENSIONS

Wingspan:	36 ft 2 in
Length:	27 ft 8 in
Height:	8 ft 2 in

WEIGHTS

Gross Weight:	3,600 lbs
Empty Weight:	1,940 lbs
Useful Load:	1,660 lbs

SPECIFICATIONS
PA-32-301T
Turbo Saratoga
1980-1984

SPEED

Top Speed at Sea Level:	205 mph
Cruise:	190 mph
Stall (landing configuration):	67 mph

TRANSITIONS

Takeoff over 50' obstacle:	1,590 ft
Ground Run:	1,110 ft
Landing over 50' obstacle:	1,725 ft
Ground roll:	732 ft

RATE OF CLIMB AT SEA LEVEL:	1,075 fpm
SERVICE CEILING:	20,000 ft

FUEL CAPACITY

Standard:	102 gal
ENGINE:	Lycoming TIO-540-S1AD
TBO:	1,800 hrs
Power:	300 hp

DIMENSIONS

Wingspan:	36 ft 2 in
Length:	28 ft 2 in
Height:	8 ft 2 in

WEIGHTS

Gross Weight:	3,600 lbs
Empty Weight:	2,003 lbs
Useful Load:	1,597 lbs

SPECIFICATIONS
PA-32R-301
Saratoga SP
1980-Present

SPEED

Top Speed at Sea Level:	188 mph
Cruise:	182 mph
Stall (landing configuration):	66 mph

TRANSITIONS

Takeoff over 50' obstacle:	1,573 ft
Ground Run:	1,013 ft
Landing over 50' obstacle:	1,612 ft
Ground Roll:	732 ft

RATE OF CLIMB AT SEA LEVEL:	1,010 fpm
SERVICE CEILING:	16,700 ft

FUEL CAPACITY

Standard:	102 gal
ENGINE:	Lycoming IO-540-K1G5D
TBO:	2,000 hrs
Power:	300 hp

DIMENSIONS

Wingspan:	36 ft 2 in
Length:	27 ft 8 in
Height:	8 ft 2 in

WEIGHTS

Gross Weight:	3,600 lbs
Empty Weight:	1,999 lbs
Useful Load:	1,601 lbs

SPECIFICATIONS

PA-32R-301T

Turbo Saratoga SP

1980-Present

SPEED

Top Speed at Sea Level:	224 mph
Cruise:	203 mph
Stall (landing configuration)	65 mph

TRANSITIONS

Takeoff over 50' obstacle:	1,420 ft
Ground Run:	960 ft
Landing over 50' obstacle:	1,725 ft
Ground Roll:	732 ft

RATE OF CLIMB AT SEA LEVEL:	1,120 fpm
SERVICE CEILING:	20,000 ft

FUEL CAPACITY

Standard:	102 gal
ENGINE:	Lycoming TIO-540-S1AD
TBO:	1,800 hrs
Power:	300 hp

DIMENSIONS

Wingspan:	36 ft 2 in
Length:	28 ft 2 in
Height:	8 ft 2 in

WEIGHTS

Gross Weight:	3,600 lbs
Empty Weight:	2,078 lbs
Useful Load:	1,522 lbs

Production Figures and Serial Numbers
PA-32 Saratogas

The following list shows the production years and total number of PA-32-301/301T/R-301/R-301T airplanes that were built, along with serial numbers.

Production Year	Total Aircraft Produced	Serial Numbers Beginning	Ending
Model PA-32-301			
1980	106	32-8006001	32-8006106
1981	99	32-8106001	32-8106099
1982	40	32-8206001	32-8206040
1983	31	32-8306001	32-8306031
1984	19	32-8406001	32-8406019
1985	21	32-8506001	32-8506021
1986	3	32-8606001	3206003
1987	n/a	3206004	n/a
Model PA-32-301T			
1980	52	32-8024001	32-8024152
1981	34	32-8124001	32-8124034
1982	10	32-8224001	32-8224010
1983	13	32-8324001	32-8324013
1984	2	32-8424001	32-8424002

Model PA-32R-301

1980	139	32R-8013001	32R-8013139
1981	122	32R-8113001	32R-8113122
1982	60	32R-8213001	32R-8213060
1983	29	32R-8313001	32R-8313029
1984	24	32R-8413001	32R-8406024
1985	16	32R-8513001	32R-8513016
1986	2	32R-8613001	3213002
1987	n/a	3213003	n/a

Model PA-32R-301T

1980	121	32R-8029001	32R-8029121
1981	114	32R-8129001	32R-8129114
1982	68	32R-8229001	32R-8229068
1983	40	32R-8329001	32R-8329040
1984	27	32R-8429001	32R-8429027
1985	20	32R-8529001	32R-8529020
1986	2	32R-8629001	3229002
1987	n/a	3229003	n/a

OBSERVATIONS

I always enjoy flying the Cherokee Sixes; they are large planes for singles, yet fly easily. The PA-32 is considerably longer than the PA-28, but I think the length is accentuated by the long nose, which houses not only the larger six-cylinder engine, but a baggage compartment as well.

Flying solo, as I often do, seemed a waste in the Six. It was sort of like flying an airliner by yourself—lots of empty space. This is why these planes have always been popular as air ambulances. There's loads of space inside, and with the optional rear door, you can put almost anything inside. On a later flight, I loaded up with all the "airport bums" and quickly noted that the plane seats six comfortably, but when you put the third person across the rear seat, it gets crowded.

Ground handling the Six is a little different from the PA-28. The nosewheel is very heavy, so the use of differential braking is a must in tight quarters. The Six handles like the heavy plane that it is.

When climbing, there is somewhat reduced forward visibility due to the long nose. Also, due to the 300 horses pulling up front, I needed to use the rudder trim, or keep my foot on the rudder, to counteract the torque. This is quite normal in a high-power single-engine airplane.

In the air, I found control pressures to be slightly heavier than the PA-28s, but this is a much heavier plane. Again, as on the ground, it handles like a heavy plane.

Stalls were tame, not unlike the PA-28-235—always with lots of warning and very gentle. With full power and solo, I was unable to get a break in the power stalls.

The flaps are not overly effective, unless you put down a full 45 degrees; then the plane comes down fast, and payoff comes quickly. You have to be on your toes for this, but it can get you into a tight field easily.

Fuel management is via the now very familiar Left-Off-Right selector found on most other Piper singles. Fuel consumption seems to be in the area of 16 to 18 gallons per hour, but this will depend on what you're doing with the plane. It's very easy to use a lot of fuel if you're taxing the 300 horses.

I flew one "T" tail, and, like many other pilots not familiar with how the Lance II flies, did not like it. It seemed the ground run was much faster, as a result of the tail having no real effect until well up in airspeed. Other than the tail difference, and the way landings and takeoffs are performed, the Lance II flies about the same as a Six. I talked with a couple of pilots who have considerable time in Lance IIs, and they say that it's just a matter of getting used to the different transition factors created by the "T" tail.

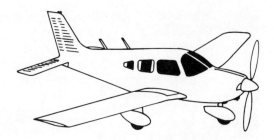

Chapter 6

The Tomahawk, Apache, and Other Indians

In 1978 Piper decided it was time to get back into the two-place trainer business. To do this, they introduced the PA-38 Tomahawk (Fig. 6-1).

A NEW-LOOK TRAINER

The Tomahawk was a new-look trainer, with no resemblance to the PA-22-108 Colt (Fig. 6-2). The little trainer had low wings, excellent visibility, wide-stance landing gear, a nicely planned panel, and a throttle guadrant that would make any airline-wishful pilot happy.

Unfortunately, this is where most of the good about the Tomahawk stops. The early PA-38s were hit with a truckload of Airworthiness Directives (ADs). Additionally, the stall and slow-flight characteristics were not forgiving, therefore not desirable for modern training. In fact, due to these stall characteristics, the PA-38 is oft-times referred to as the ''Traumahawk.''

Changes

1981: ☐ The Tomahawk II begins deliveries.

New Prices

1978:	☐ PA-38 Tomahawk	$15,820
1981:	☐ PA-38 Tomahawk II	$22,090

Fig. 6-1. The Piper Tomahawk is a very different airplane from the old Cubs. New smooth lines and excellent visibility make it a very modern trainer.

Warning

Many of the ADs directed toward the PA-38 are quite expensive to comply with. Watch out for these expensive bombshells when looking with an eye towards purchase. If they haven't been complied with, look elsewhere. There are lots of Tomahawks for sale. The Tomahawk II is essentially the same plane as the original Tomahawk, but built so as not to be subject to the expensive early ADs.

78-22-1: Modify clearance between the rudder leading edge and the fin trailing edge.

78-23-4: Replace missing rivets in the rear-wing spar-to-fuselage attach fitting.

79-8-2: Inspect/replace stabilizer pulley mounting bracket, and loose stabilizer-to-fin and fin-to-fuselage attach bolts.

81-23-7: *Expensive!*—Mandates the replacement of the engine mount prior to 1000 hours of operation, to preclude possible landing gear failure.

82-2-1: Inspect/repair aileron balance-weight rib flanges.

78-25-2, 80-6-5, 80-25-2, 80-25-7, and **81-18-4** are all engine-related ADs.

SPECIFICATIONS

PA-38-112

Tomahawk & Tomahawk II

1978-Present

SPEED

Top Speed at Sea Level:	125 mph
Cruise:	124 mph
Stall (landing configuration):	56 mph

TRANSITIONS

Takeoff over 50' obstacle:	1,460 ft
Ground Run:	820 ft
Landing over 50' obstacle:	1,465 ft
Ground Roll:	525 ft

RATE OF CLIMB AT SEA LEVEL: 718 fpm

SERVICE CEILING: 13,000 ft

FUEL CAPACITY

Standard:	32 gal

ENGINE: Lycoming O-235-L2C

TBO:	2,000 hrs
Power:	112 hp

DIMENSIONS

Wingspan:	34 ft 0 in
Wing Area:	124.7 sq ft
Length:	22 ft
Height:	9 ft 1 in
Wing Loading:	13.4 lb/sq ft
Power Loading:	14.9 lb/hp

WEIGHTS

Gross Weight:	1,670 lbs
Empty Weight:	1,109 lbs
Useful Load:	561 lbs
Baggage Allowance:	100 lbs

Fig. 6-2. The Colt, the last of the tube-and-fabric trainers. Until the Tomahawk was introduced, Piper built no true trainers after Colt production stopped.

Production Figures and Serial Numbers
Model PA-38-112

The following list shows the production years and total number of PA-38 airplanes built, along with serial numbers.

Production Year	Total Aircraft Produced	Serial Numbers Beginning	Ending
1978	821	38-78A0001	38-78A0821
1979	1,179	38-79A0001	38-79A1179
1980	189	38-80A0001	38-80A0189
1981	173	38-81A0001	38-81A0173
1982	122	38-82A0001	38-82A0122

Observations

I've flown many types of trainers over the years. Some were slow, some were fast, some would keep you on your toes. The Tomahawk is the latter . . . it will keep you on your toes.

For a trainer, the PA-38 provides many surprises for the student. Perhaps the most startling of these ''training'' characteristics is the unpredictable stalls.

When a student stalls the Tomahawk, he will know *for sure* he has stalled. There is no "slight mush" here. Stalls are full break, and to any direction. Of course, if speeds and coordination are as they should be, then you have no stalls.

Some flight instructors feel these stall characteristics are not good for the student to have to contend with. Other instructors feel this vigorous training makes for good pilots. I subscribe to the latter theory.

Other points of note:

The Tomahawk has excellent visibility and a wide cabin that is fairly comfortable.

The use of noise-reducing headphones is recommended to keep headaches and buzzing ears to a minimum—the cabin is too noisy.

The tail shakes during a stall. In fact, it shakes so much it might scare you.

OTHER INDIANS

As with each of the Indians already discussed, the remaining Indians were each built to provide a particular service. Descriptions of most of these models follow.

Apache and Aztec

The PA-23 Apache was actually the first Piper Indian. It was also Piper's first light twin and was introduced in 1954 (Fig. 6-3).

Unfortunately, the Apache was a poor performer and looked ungainly.

Several changes were made over the years to the PA-23 until

Fig. 6-3. The Piper Apache is often referred to as the "Flying Potato," due to its shape.

Fig. 6-4. The Piper Aztec, although still considered a PA-23, is quite different from the Apache.

the basic airplane no longer looked like an Apache. These changes included seating, additional power, and airframe lines changes. The Apaches last year in production was 1963.

Due to the changes made to the Apache, Piper brought about a name change. In 1960, the first sales were made of the Aztec, also known as a PA-23. Appearance-wise the new plane was sleeker and had a swept fin (Fig. 6-4). The Aztec was the evolutionary result of the Apache changes.

APACHE

ENGINE:	150-hp Lycoming O-320	
PERFORMANCE		
Cruise Speed:	170	mph
Takeoff over 50' obstacle:	1,600	ft
Landing over 50' obstacle:	1,360	ft
Rate of Climb:	1,350	fpm
Service Ceiling:	20,000	ft
WEIGHT		
Gross:	3,500	lbs
Useful Load:	1,320	lbs
SEATS:	4	

AZTEC

ENGINE:	250-hp Lycoming IO-540	
PERFORMANCE		
Cruise Speed:	205	mph
Takeoff over 50' obstacle:	1,100	ft
Landing over 50' obstacle:	1,260	ft
Rate of Climb:	1,650	fpm
Service Ceiling:	22,500	ft
WEIGHT		
Gross:	4,800	lbs
Useful Load:	1,900	lbs
SEATS:	6	

Twin Comanche

In 1963, Piper introduced a new light twin, the PA-30. The Twin Comanche was somewhat similar in appearance to the Cessna 310, but was much smaller and lighter (Fig. 6-5).

In 1970, Piper improved the PA-30 by adding counter-rotating engines, thus eliminating the "critical engine" effect. Although the name "Twin Comanche" remained unchanged, the model number was changed to PA-39.

Production of the Twin Comanche ended in 1972.

TWIN COMANCHE

ENGINE:	160-hp Lycoming IO-320	
PERFORMANCE		
Cruise Speed:	198	mph
Takeoff over 50' obstacle:	1,530	ft
Landing over 50' obstacle:	1,870	ft
Rate of Climb:	1,460	fpm
Service Ceiling:	20,000	ft
WEIGHT		
Gross:	3,600	lbs
Useful Load:	1,330	lbs
SEATS:	6	
(early models)	4	

Fig. 6-5. The small, sleek Twin Comanche. It can be found as a PA-30 or a PA-39; the latter has counter-rotating engines.

Seneca

Piper first introduced the Seneca series of light twins in 1972 as the PA-34 Model 200. It seated six passengers and was powered with fuel-injected Lycoming 200-hp engines. The Seneca was basically a Cherokee Six made twin.

In 1975 the Seneca II Model 200T with a Teledyne Continental turbocharged engine was brought out. This version seated seven.

The current models of the Seneca are the PA-34-220/220T Seneca IIIs.

SENECA

ENGINE:	220-hp Teledyne Continental TSIO-360	
PERFORMANCE		
Cruise Speed:	193	kts
Takeoff over 50' obstacle:	1,210	ft
Landing over 50' obstacle:	2,160	ft
Rate of Climb:	1,400	fpm
Service Ceiling:	25,000	ft
Maximum Range:	990	nm
WEIGHT		
Gross:	4,773	lbs
Useful Load:	1,921	lbs
SEATS:	6-7	
SUGGESTED PRICE (new):	$194,900	

Fig. 6-6. The Piper Seneca III, an outgrowth of the PA-32. (courtesy Piper Aircraft)

Navajo

From the smaller twins to the largest cabin-class airplanes, Piper has built airplanes to suit every market and pocketbook.

The cabin-class twins are primarily business aircraft, although some are used in air transport with small commuter airlines, and for package delivery.

The PA-31P-425 pressurized Navajo was in production from 1970 through 1977 and was quite popular with commuter airlines.

NAVAJO		
ENGINES:	425-hp Lycoming TIGO-541-E1A	
TBO:	1,200	hrs
PERFORMANCE		
Cruise Speed:	218	kts
Maximum Speed:	241	kts
Stall, normal:	72	kts
Takeoff over 50' obstacle:	2,200	ft
Ground Run:	1,440	ft
Landing over 50' obstacle:	2,700	ft
Ground Roll:	1,370	ft
Rate of Climb:	1,740	fpm
Service Ceiling:	29,000	ft
WEIGHT		
Gross:	7,800	lbs
Empty:	5,000	lbs
SEATS:	8	
DIMENSIONS		
Length	34 ft	5 in
Height	13 ft	3 in
Wingspan	40 ft	7 in
SUGGESTED PRICE (new, 1977)	$272,450	

Chieftain

The PA-31-350 Chieftain has been in production since 1973, and also utilizes the PA-31 airframe and standard reciprocating engines (Fig. 6-7).

CHIEFTAIN

ENGINE:	350-hp Lycoming TIO-540
PERFORMANCE	
Cruise Speed:	221 kts
Takeoff over 50' obstacle:	2,510 ft
Landing over 50' obstacle:	1,880 ft
Rate of Climb:	1,120 fpm
Service Ceiling:	24,000 ft
Maximum Range:	1,275 nm
WEIGHT	
Gross:	7,045 lbs
Useful Load:	2,726 lbs
SEATS:	6-10
SUGGESTED PRICE (new):	$470,900

Cheyenne

Piper's Cheyenne is the top of the line. Now known as the PA-42, the Cheyenne is the culmination of a series of aircraft based on the PA-31 airframe (Figs. 6-8, 6-9).

The Cheyenne series is equipped with turbine engines of varying shaft horsepower, and commensurate varying performance. 1987 prices start in excess of $2.7 million.

CHEYENNE IA

ENGINE:	500-shp Pratt & Whitney PT6A-11
PERFORMANCE	
Cruise Speed:	261 kts
Takeoff over 50' obstacle:	2,444 ft
Landing over 50' obstacle:	2,263 ft
Rate of Climb:	1,750 fpm
Service Ceiling:	28,200 ft
Maximum Range:	1,300 nm
WEIGHT	
Gross:	8,750 lbs
Useful Load:	3,646 lbs
SEATS:	7
SUGGESTED PRICE (new):	$1,042,350

Fig. 6-7. Piper Chieftain.

Fig. 6-8. The Cheyenne I-A. (courtesy Piper Aircraft)

Fig. 6-9. The Cheyenne III-A. (courtesy Piper Aircraft)

CHEYENNE IIXL

ENGINE: 620-shp Pratt & Whitney PT6A-135

PERFORMANCE

Cruise Speed:	275	kts
Takeoff over 50′ obstacle:	2,446	ft
Landing over 50′ obstacle:	1,773	ft
Rate of Climb:	1,750	fpm
Service Ceiling:	32,400	ft
Maximum Range:	1,280	nm

WEIGHT

Gross:	9,540	lbs
Useful Load:	4,053	lbs

SEATS: 7-8

SUGGESTED PRICE (new): $1,483,08

CHEYENNE IIIA

ENGINE: 720-shp Pratt & Whitney PT6A-61

PERFORMANCE

Cruise Speed:	305	kts
Takeoff over 50′ obstacle:	2,280	ft
Landing over 50′ obstacle:	3,043	ft
Rate of Climb:	2,380	fpm
Service Ceiling:	38,840	ft
Maximum Range:	2,270	nm

WEIGHT

Gross:	11,285	lbs
Useful Load:	4,448	lbs

SEATS: 9-11

SUGGESTED PRICE (new): $1,995,00

CHEYENNE 400LS

ENGINE: 1000-shp Garrett TPE 331-14

PERFORMANCE

Cruise Speed:	351	kts
Takeoff over 50′ obstacle:	1,930	ft
Landing over 50′ obstacle:	2,280	ft
Rate of Climb:	3,242	fpm
Service Ceiling:	41,000	ft
Maximum Range:	2,176	nm

WEIGHT

Gross:	12,135	lbs
Useful Load:	4,589	lbs

SEATS: 9

SUGGESTED PRICE (new): $2,375,00

Mojave

The most recent introduction using the PA-31 airframe is the PA-31P-350 Mojave, a pressurized airframe with piston engines.

MOJAVE		
ENGINE:	350-hp Lycoming TIO-540	
PERFORMANCE		
Cruise Speed:	235	kts
Takeoff over 50' obstacle:	3,035	ft
Landing over 50' obstacle:	2,300	ft
Rate of Climb:	1,220	fpm
Service Ceiling:	26,500	ft
Maximum Range:	1,221	nm
WEIGHT		
Gross:	7,245	lbs
Useful Load:	2,181	lbs
SEATS:	7	
SUGGESTED PRICE (new):	$657,000	

Aerostar

In 1969, the first Aerostar was produced. Originally known as the Aerostar 600, it was a nonpressurized twin powered with 290-hp Lycoming TIO-540s.

Since the first Aerostars were built there have been many improvements, the most significant being pressurization. The 601P and 602P were both pressurized.

The most recent version of the Aerostar is the PA-60 Model 700P (Fig. 6-10). Piper claimed this aircraft would give propjet performance at half the cost, but production is currently suspended.

Fig. 6-10. Piper Aerostar. (courtesy Piper Aircraft)

AEROSTAR 700P

ENGINE:	350-hp Lycoming TIO-540	
PERFORMANCE		
Cruise Speed:	230	kts
Takeoff over 50' obstacle:	3,080	ft
Landing over 50' obstacle:	2,140	ft
Rate of Climb:	1,840	fpm
Service Ceiling:	25,000	ft
Maximum Range:	1,160	nm
WEIGHT		
Gross:	6,315	lbs
Useful Load:	2,081	lbs
SEATS:	6	
SUGGESTED PRICE (new):	$514,280	

Malibu

The newest addition to Piper's cabin-class field is the Malibu. Known as Model PA-46, the Malibu is a single-engine aircraft (Figs. 6-11, 6-12).

Piper advertises this pressurized single as the "fastest production piston single you can find."

MALIBU

ENGINE:	310-hp Teledyne Continental TSIO-520	
PERFORMANCE		
Cruise Speed:	187	kts
Takeoff over 50' obstacle:	2,025	ft
Landing over 50' obstacle:	1,800	ft
Rate of Climb:	1,143	fpm
Service Ceiling:	25,000	ft
Maximum Range:	1,555	nm
WEIGHT		
Gross:	4,118	lbs
Useful Load:	1,652	lbs
SEATS:	6	
SUGGESTED PRICE (new):	$300,000	

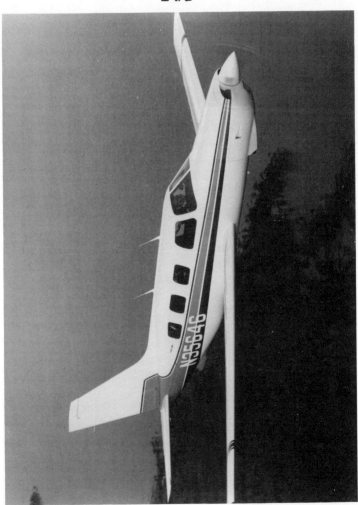

Fig. 6-11. Piper Malibu, the swankiest single-engine airplane on the market. (courtesy Piper Aircraft)

Fig. 6-12. The very plush interior of the Malibu—true "cabin class" in a single. (courtesy Piper Aircraft)

Seminole

The PA-44 Seminole twin was produced between 1979 and 1982, with a Turbo version available in 1981.

SEMINOLE

ENGINE:	180-hp Lycoming 0360-E1A6D		
PERFORMANCE			
Cruise Speed:	159	kts	
Maximum Speed:	166	kts	
Stall, normal:	54	kts	
Takeoff over 50' obstacle:	1,400	ft	
Ground Run:	880	ft	
Landing over 50' obstacle:	1,400	ft	
Ground Roll:	590	ft	
Rate of Climb:	1,340	fpm	
Service Ceiling:	17,100	ft	
WEIGHT			
Gross:	3,800	lbs	
Empty	2,406	lbs	
SEATS:	4		
DIMENSIONS			
Length	34	ft	5 in
Height	13	ft	3 in
Wingspan	40	ft	7 in

Fig. 6-13. The hard-working Pawnee crop duster. (courtesy Piper Aircraft)

Pawnee

Possibly the hardest working Indian is the Pawnee, a crop-dusting aircraft first built in 1960.

Earlier models were known as PA-25s and were powered with 235-hp Lycoming engines.

In 1973, the Pawnee Brave (PA-36) was introduced with an optional 400-hp engine (Fig. 6-13).

PAWNEE BRAVE

ENGINES:	375-hp Lycoming
PERFORMANCE	
Cruise Speed:	102 kts
Takeoff over 50' obstacle:	1,765 ft
Landing over 50' obstacle:	1,200 ft
Rate of Climb:	550 fpm
WEIGHT	
Gross:	4,800 lbs
Useful Load:	2,280 lbs
SUGGESTED PRICE (new):	$119,500

Chapter 7
Indian Engines

Numerous models of engines can be found powering the various Piper Indian airplanes. They will vary in power (112 to 300 horsepower), manufacture (Continental or Lycoming), and even aspiration (carbureted, fuel-injected, or turbocharged).

ENGINE SPECIFICATIONS
The following engine specifications indicate the particular aircraft models in which the engines were originally installed. This may or may not mean the same type engine is still in an older-model aircraft. Changes do get made for one reason or another.

In reading the specifications charts, notice that the engine model numbers are in three parts. These parts completely describe a particular model. For example, the O-320-L2C engine decodes as:

O—Horizontally opposed engine. Other designators you will see for Piper Indians are IO (fuel-injected, horizontally opposed) and TIO (turbocharged, fuel-injected, horizontally opposed).
320—The cubic-inch displacement of the engine.
L2C—The particular version of the basic engine.

Versions may differ in head design, accessory box design, and ignition.

Engines with few basic differences have been grouped together.

LYCOMING O-235-L2C/N2C

USED ON:	PA-38-112 Tomahawk	
HORSEPOWER:	118	
RATED RPM		
Full Throttle:	2,800	
65%:	2,400	
FUEL CONSUMPTION:	6.1	gph
DIMENSIONS		
Bore:	4.375	in
Stroke:	3.875	in
Weight:	252	lbs
Height:	22.40	in
Width:	32.00	in
Length:	29.56	in
CYLINDERS:	4	
DISPLACEMENT:	233.3	cu in
COMPRESSION RATIO:	8.5:1	
OIL CAPACITY:	6	qts
FUEL GRADE:	100/100LL	

LYCOMING O-320-A2B/E2A/E3D

USED ON:	PA-28-150	Cherokee 150	
	PA-28-140	Cherokee 140	
		Cruiser	
	PA-28-151	Warrior	
HORSEPOWER:	150		
RATED RPM			
Full Throttle:	2,700		
75%:	2,450		
65%:	2,350		
FUEL CONSUMPTION			
75% Throttle:	10.0		gph
65%:	8.8		gph
60%:	8.2		gph
DIMENSIONS			
Bore:	5.125		in
Stroke:	3.875		in
Displacement:	319.8		cu in
Compression Ratio:	7.0:1		
Weight:	272		lbs
Height:	22.99		in
Width:	32.24		in
Length:	29.56		in
CYLINDERS:	4		
OIL CAPACITY:	8		qts
FUEL GRADE:	80/87		

LYCOMING O-320-B2B/D2A/D3G

USED ON:	PA-28-160	Cherokee 160
	PA-28-161	Warrior II
HORSEPOWER:	160	
RATED RPM		
Full Throttle:	2,700	
75%:	2,450	
65%:	2,350	
FUEL CONSUMPTION		
75% Throttle:	10.0	gph
65%:	8.8	gph
60%:	8.2	gph
DIMENSIONS		
Bore:	5.125	in
Stroke:	3.876	in
Displacement:	319.8	cu in
Compression Ratio:	8.5:1	
Weight:	278	lbs
Height:	22.99	in
Width:	32.24	in
Length:	29.56	in
CYLINDERS:	4	
OIL CAPACITY:	8	qts
FUEL GRADE:	100/100LL	

LYCOMING O-320-E2A/E3D
(Fig. 7-1)

USED ON:	PA-28-140 Cherokee 140	
HORSEPOWER:	140	
RATED RPM		
Full Throttle:	2,450	
75%:	2,300	
65%:	2,150	
FUEL CONSUMPTION		
75% Throttle:	8.8	gph
65%:	8.2	gph
60%:	7.8	gph
DIMENSIONS		
Bore:	5.125	in
Stroke:	3.875	in
Displacement:	319.8	cu in
Compression Ratio:	7.0:1	
Weight:	272	lbs
Height:	22.99	in
Width:	32.24	in
Length:	29.56	in
CYLINDERS:	4	
OIL CAPACITY:	8	qts
FUEL GRADE:	80/87	

Fig. 7-1. The Lycoming 0-320 engine is found in many versions of the PA-28s. (courtesy AVCO Lycoming)

LYCOMING O-360-A1A/A3A/A4A/A4M/A4N (Fig. 7-2)

USED ON:	PA-28-180	Cherokee 180
		Challenger
		Archer
	PA-28-181	Archer II
HORSEPOWER:	180	
RATED RPM		
Full Throttle:	2,700	
75%:	2,450	
65%:	2,350	
FUEL CONSUMPTION		
75% Throttle:	10.5	gph
65%:	9.0	gph
60%:	8.5	gph
DIMENSIONS		
Bore:	5.125	in
Stroke:	4.375	in
Displacement:	361	cu in
Compression Ratio:	8.5:1	
Weight:	285	lbs
Height:	24.59	in
Width:	33.37	in
Length:	29.56	in
CYLINDERS:	4	
OIL CAPACITY:	8	qts
FUEL GRADE:	100/100LL	

LYCOMING IO-360-B1E

USED ON:	PA-28R-180	Arrow 180
HORSEPOWER:	180	
RATED RPM		
Full Throttle:	2,700	
75%:	2,450	
65%:	2,350	
FUEL CONSUMPTION		
75% Throttle:	11.0	gph
65%:	8.5	gph
60%:	8.0	gph
DIMENSIONS		
Bore:	5.125	in
Stroke:	4.375	in
Displacement:	361	cu in
Compression Ratio:	8.5:1	
Weight:	291	lbs
Height:	20.70	in
Width:	33.37	in
Length:	32.39	in
CYLINDERS:	4	
OIL CAPACITY:	8	qts
FUEL GRADE:	100/100LL	

Fig. 7-2. Lycoming O-360 engines power the 180-hp PA-28s. This particular engine is an IO-360, equipped with fuel injection. (courtesy AVCO Lycoming)

LYCOMING IO-360-C1C/C1C6

USED ON:	PA-28RB-200	Arrow	
	PA-28R-201	Arrow II	
	PA-28RT-201	Arrow III	
		Arrow IV	
HORSEPOWER:		200	
RATED RPM			
Full Throttle:		2,700	
75%:		2,450	
65%:		2,350	
FUEL CONSUMPTION			
75% Throttle:		12.3	gph
65%:		9.5	gph
60%:		9.0	gph
DIMENSIONS			
Bore:		5.125	in
Stroke:		4.375	in
Displacement:		361	cu in
Compression Ratio:		8.7:1	
Weight:		328	lbs
Height:		19.48	in
Width:		34.25	in
Length:		33.65	in
CYLINDERS:		4	
OIL CAPACITY:		8	qts
FUEL GRADE:		100/100LL	

CONTINENTAL TSIO-360-F/FB

USED ON:	PA-28-201T	Turbo Dakota	
	PA-28R-201T	Turbo Arrow III	
	PA-28RT-201T	Turbo Arrow IV	
HORSEPOWER:		200	
RATED RPM			
Full Throttle:		2,575	
DIMENSIONS			
Bore:		4.44	in
Stroke:		3.87	in
Displacement:		360	cu in
Compression Ratio:		7.5:1	
Weight:		359	lbs
Height:		26.44	in
Width:		31.30	in
Length:		56.58	in
CYLINDERS:		6	
FUEL GRADE:		100/130	

LYCOMING O-540-A1A5

USED ON:	PA-24-250 Comanche 250
HORSEPOWER:	250
RATED RPM	
Full Throttle:	2,575
75%:	2,350
60%:	2,220
FUEL CONSUMPTION	
75% Throttle:	16.3 gph
65%:	13.5 gph
60%:	12.5 gph
DIMENSIONS	
Bore:	5.125 in
Stroke:	4.375 in
Displacement:	541.5 cu in
Compression Ratio:	8.5:1
Weight:	405 lbs
Height:	24.56 in
Width:	33.37 in
Length:	37.22 in
CYLINDERS:	6
OIL CAPACITY:	1275 qts
FUEL GRADE:	91/96 or 100/130

LYCOMING O-540-B1B5/B2B5/B4B5

USED ON:	PA-28-235 Cherokee 235 Pathfinder Charger
HORSEPOWER:	235
RATED RPM	
Full Throttle:	2,575
75%:	2,350
65%:	2,200
FUEL CONSUMPTION	
75% Throttle:	15.5 gph
65%:	13.0 gph
60%:	12.2 gph
DIMENSIONS	
Bore:	5.125 in
Stroke:	4.375 in
Displacement:	541 cu in
Compression Ratio:	7.2:1
Weight:	395 lbs
Height:	24.56 in
Width:	33.37 in
Length:	38.42 in
CYLINDERS:	6
OIL CAPACITY:	12 qts
FUEL GRADE:	80/87

LYCOMING O-540-E4A5/E4B5

USED ON:	PA-24-260 Comanche 260	
	PA-28-260 Cherokee Six	
HORSEPOWER:	260	
RATED RPM		
Full Throttle:	2,700	
75%:	2,450	
65%:	2,350	
FUEL CONSUMPTION		
75% Throttle:	19.0	gph
65%:	15.5	gph
60%:	14.0	gph
DIMENSIONS		
Bore:	5.125	in
Stroke:	4.375	in
Displacement:	541.5	cu in
Compression Ratio:	8.5:1	
Weight:	397	lbs
Height:	24.56	in
Width:	33.37	in
Length:	38.42	in
CYLINDERS:	6	
OIL CAPACITY:	12	qts
FUEL GRADE:	91/96	

LYCOMING IO-540-D4A5

USED ON:	PA-24-260 Comanche	
HORSEPOWER:	260	
RATED RPM		
Full Throttle:	2,700	
75%:	2,450	
60%:	2,350	
FUEL CONSUMPTION		
75% Throttle:	15.0	gph
65%:	12.5	gph
60%:	12.0	gph
DIMENSIONS		
Bore:	5.125	in
Stroke:	4.375	in
Displacement:	541.5	cu in
Compression Ratio:	8.5:1	
Weight:	410	lbs
Height:	24.56	in
Width:	33.37	in
Length:	38.42	in
CYLINDERS:	6	
OIL CAPACITY:	12	qts
FUEL GRADE:	91/96 or 100/130	

LYCOMING O-540-J3A5D

USED ON:	PA-28-236 Dakota
HORSEPOWER:	235
RATED RPM	
Full Throttle:	2,400
75%:	2,200
65%:	2,000
FUEL CONSUMPTION	
75% Throttle:	14.0 gph
65%:	10.8 gph
DIMENSIONS	
Bore:	5.125 in
Stroke:	4.375 in
Displacement:	541.5 cu in
Compression Ratio:	8.5:1
Weight:	388 lbs
Height:	24.56 in
Width:	33.37 in
Length:	38.93 in
CYLINDERS:	6
OIL CAPACITY:	12 qts
FUEL GRADE:	100/100LL

LYCOMING IO-540-K1A5D/K1G5/K1G5D

USED ON:	PA-32-300	Cherokee Six
	PA-32R-300	Lance
		Lance II
	PA-32-301	Saratoga
	PA-32R-301	Saratoga SP
HORSEPOWER:		300
RATED RPM		
Full Throttle:		2,700
75%:		2,450
60%:		2,350
FUEL CONSUMPTION		
75% Throttle:		18.0 gph
65%:		13.8 gph
60%:		12.8 gph
DIMENSIONS		
Bore:		5.125 in
Stroke:		4.375 in
Displacement:		541.5 cu in
Compression Ratio:		8.7:1
Weight:		470 lbs
Height:		19.60 in
Width:		34.25 in
Length:		38.93 in
CYLINDERS:		6
OIL CAPACITY:		12 qts
FUEL GRADE:		100/130

LYCOMING IO-540-N1A5

USED ON:	PA-24-260 Comanche Turbo 260
HORSEPOWER:	260
RATED RPM	
Full Throttle:	2,700
75%:	2,400
60%:	2,200
FUEL CONSUMPTION	
75% Throttle:	16.0 gph
65%:	15.4 gph
60%:	12.6 gph
DIMENSIONS	
Bore:	5.125 in
Stroke:	4.375 in
Displacement:	541.5 cu in
Compression Ratio:	7.3:1
Weight:	410 lbs
Height:	24.56 in
Width:	36.02 in
Length:	39.56 in
CYLINDERS:	6
OIL CAPACITY:	12 qts
FUEL GRADE:	100/130

LYCOMING TIO-540-S1AD
(Fig. 7-3)

USED ON:	PA-32RT-300T Turbo Lance II
	PA-32-301T Turbo Saratoga
	PA-32R-301T Turbo Saratoga SP
HORSEPOWER:	300
RATED RPM	
Full Throttle:	2,700
75%:	2,400
60%:	2,200
FUEL CONSUMPTION	
75% Throttle:	19.0 gph
65%:	14.8 gph
60%:	13.6 gph
DIMENSIONS	
Bore:	5.125 in
Stroke:	4.375 in
Displacement:	541.5 cu in
Compression Ratio:	7.3:1
Weight:	533 lbs
Height:	26.28 in
Width:	36.02 in
Length:	39.56 in
CYLINDERS:	6
OIL CAPACITY:	12 qts
FUEL GRADE:	100/130

Fig. 7-3. The 540 engine is found in many Piper airplanes. This is a TIO version. (courtesy Lycoming)

Fig. 7-4. Notice the turbocharger on the rear of this TO-360 engine. (courtesy AVCO Lycoming)

123

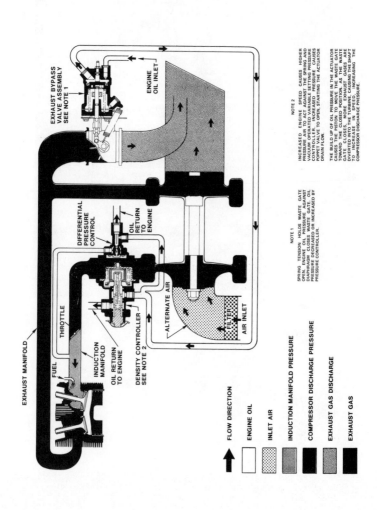

Fig. 7-5. Theory of operation for a turbocharging system.

EXHAUST BYPASS VALVE ASSEMBLY SEE NOTE 1

ENGINE OIL INLET

DIFFERENTIAL PRESSURE CONTROL

OIL RETURN TO ENGINE

EXHAUST MANIFOLD

THROTTLE

FUEL

INDUCTION MANIFOLD

OIL RETURN TO ENGINE

DENSITY CONTROLLER SEE NOTE 2

ALTERNATE AIR

FILTER

AIR INLET

NOTE 1

SPRING TENSION HOLDS WASTE GATE OPEN, ENGINE OIL PRESSURE AGAINST DIAPHRAGM CLOSES WASTE GATE. OIL PRESSURE DECREASED OR INCREASED BY PRESSURE CONTROLLER.

NOTE 2

INCREASED ENGINE SPEED CAUSES HIGHER PRESSURE AIR TO ACT AGAINST THE SPRING AND VACUUM OPERATED VARIABLE SETTING PRESSURE CONTROLLER. INCREASED PRESSURE CAUSES POPPET VALVE TO OPEN, STARTING THE ACTUATOR DRAIN FLOW.

THE BUILD UP OF OIL PRESSURE IN THE ACTUATOR CAUSES THE PISTON TO MOVE THE WASTE GATE TOWARD THE CLOSED POSITION. AS THE WASTE GATE CLOSES, MORE EXHAUST GASES ARE DIVERTED INTO THE TURBINE, CAUSING THE SHAFT TO INCREASE IN SPEED, INCREASING THE COMPRESSOR DISCHARGE PRESSURE.

FLOW DIRECTION

ENGINE OIL

INLET AIR

INDUCTION MANIFOLD PRESSURE

COMPRESSOR DISCHARGE PRESSURE

EXHAUST GAS DISCHARGE

EXHAUST GAS

124

TURBOCHARGING

The normally aspirated (non-supercharged) aircraft engine fails miserably in performance at high altitudes. This is due to thinner air. The higher the altitude, the thinner the air.

In order to force acceptable performance from an aircraft engine when flying at high altitudes, some means of pumping more air into the combustion chambers must be provided, thus compensating for the thinner air. Today's popular method for accomplishing this task is the turbocharger.

The turbocharger is a turbine wheel attached via a common shaft to a compressor wheel, each within separate housings. Exhaust gases from the engine pass through the turbine wheel housing, causing the wheel to turn, which causes the compressor wheel in the other housing to turn (Fig. 7-4).

As the compressor wheel turns, it draws air in from the outside, compresses it, and delivers it to the combustion chambers. By compressing the air, it is possible to equal sea level densities, allowing for efficient combustion and more power production (Fig. 7-5).

The end result of turbocharging is efficient engine operation from sea level to better than 20,000 feet. This high-altitude cruise allows for "above-the-weather" operations.

Chapter 8

ADs

Perfection has been one of man's goals. This is particularly true in aircraft manufacture. Unfortunately, perfection is rarely achieved . . . no airplane is perfect in design or manufacture.

When a defect in design or manufacture is discovered, corrective action must be taken to provide for continued flying safety. This corrective action may be an inspection of a known problem area to prevent a small discrepancy from progressing into a major issue, additional service (i.e., extra lubrication of a part), or repair of the problem.

THE AIRWORTHINESS DIRECTIVE

In all cases, the object of corrective action is to ensure the integrity of the aircraft, thus promoting flight safety. The FAA requires this corrective action in the form of the Airworthiness Directive (AD).

ADs are described in FAR Part 39, and must be compiled with. The AD may vary from a simple one-time inspection to a major modification of the airframe/engine. Some ADs are serious enough to cause the immediate grounding of an aircraft until compliance is met.

ADs requiring only inspections are generally inexpensive to comply with. Those involving extensive engine or airframe modifications/repairs can be extremely expensive to comply with.

ADs are not normally handled like automobile "recalls" with the manufacturer being responsible for the costs involved. Many ADs don't appear until many years after an airplane was built. The manufacturers will sometimes offer the parts/labor free of charge, but don't count on it. Even though ADs correct deficient design or poor quality control of parts or workmanship, AD compliance is routinely paid for by the airplane owner. There is no large consumer voice involving aircraft manufacturer responsibility.

Notice of an AD will be placed in the Federal Register by the FAA, and a copy will be sent by mail to all registered owners of the aircraft type concerned. In an emergency, the AD information will be sent by telegram to all registered owners.

Either way, the AD's purpose is to ensure the integrity of your flying machine, and your safety.

The records of AD compliance became a part of the aircraft's logbooks. When looking at an airplane with purchase in mind, check for AD compliance.

AD LIST

The Piper all-metal aircraft, like all other airplanes, have their share of ADs to be compiled with. *The following AD list should not be considered complete.* It is an *abbreviated* guide to assist the owner, would-be owner, or pilot in checking for AD compliance. The ADs are listed by model. The abbreviation "IAW" means "in accordance with."

For a complete check of ADs on a particular airplane, see your mechanic, or contact the Aircraft Owners and Pilots Association (AOPA). The latter will provide a list of ADs for an aircraft (by serial number) for a small fee. This type of search is highly accurate, and well worth the money spent.

Model PA-28 (All except PA-28-236)
Model PA-32-260 (All)
Model PA-32-300 (through S/N 32-7840202)

87-8-8: Inspect lower spar caps, and upper skins (Note: At press time this AD had been rescinded. However, due to structural failures occurring subsequent to the rescission, the AD might be reinstated. Inspection and/or repair can cost $1500-$4000 or more.) For the latest information contact:

Airframe Branch
Atlanta Aircraft Certification Office
FAA
Suite 210
1669 Phoenix Parkway
Atlanta, GA 30349
(404) 991-2910

Model PA-28 Cherokee (140/150/160/180/235)

62-19-3: Replace the propeller bolts (all through 28-365).

62-26-6: Inspect every 50 hours or rework the exhaust system (all through 28-707).

63-23-2: Inspect/replace the exhaust valves, Lycoming 0-320 engines, every 500 hours of operation. (*Note:* If the 7/16-inch valves have been replaced with 1/2-inch-diameter valves, the AD no longer applies, and the TBO is raised to 2000 hours.)

64-6-6: Inspect/replace the control wheel (all through 28-868).

64-16-5: Modify the fuel pump on Lycoming engines.

65-6-6: Modify the sending unit of the fuel gauge (all).

66-20-4: Replace the oil filter studs on Lycoming engines.

66-20-5: Modify the propeller spinner (Ser. 28-1761 through 28-3533).

67-12-6: Inspect certain tubing for corrosion (all).

67-20-4: Replace the nosewheel torque links.

67-26-2: Rework the fuel lines and selector valves.

67-26-3: Each 200 hours, inspect the fuel system.

69-9-3: Tachometer must be marked with red from 2150 through 2350 rpm to prevent propeller vibration and destruction.

69-15-1: Inspect/replace the control wheel pin.

69-22-2: Inspect the control wheel with dye penetrant.

70-15-18: Rework autopilot system.

70-16-5: Inspect each 100 hours or replace the muffler (all).

70-18-5: Replace landing gear attach bolts.

70-22-2: Rework the fuel selector valve (235 models only).

70-26-4: Inspect the stabilator tube with dye penetrant.

71-14-6: Inspect the magneto filters (all).

71-21-8: Replace the fuel selector cover (all except 235).

72-8-6: Inspect the nosewheel torque links (all).

72-14-7: Every 100 hours, inspect the stabilator attachment fittings.

72-17-5: Modify electric trim control system (if equipped).

72-24-2: Attach throttle control placard (140 through 160).

73-23-1: Inspect piston pins on Lycoming engines.

74-9-4: Move the rear-seat attachment points (all).

74-13-4: Install a new end bearing in the throttle control cable (140).

74-18-6: Replace the fuel quantity placards on 1973/74 235 models.

74-19-1: Inspect the lightening hole is the wing spars on 235 models.

74-24-13: Replace certain altimeters.

74-26-7: Install/replace the "Spins Prohibited" placard on 1973 through 1975 180 models.

74-26-9: Inspect certain Bendix magnetos.

75-8-3: Rework fuel drain valves (all except 235).

75-8-9: Replace the oil pump shaft and impeller on Lycoming engines.

75-24-2: Inspect/replace rear-seat legs and attachments (all).

76-18-4: Inspect the fuel selector valve (235).

76-25-6: Replace the oil radiator hose (140 only through 7125471).

77-1-1: Check system and placard fuel quantity gauges (all).

77-12-1: Replace fuel system parts or inspect every 50 hours of operation (235).

77-12-6: Replace certain Hartzell propeller blades.

77-23-3: Check/replace engine and propeller control rod/cable ends.

78-23-1: Replace fuel drain cover door or inspect every 100 hours (235).

79-7-2: Replace certain batteries.

79-13-8: Replace certain Airborne dry air pumps.

79-18-5: Replace the ELT battery.

79-22-2: Modify the fuel tank vents (all).

79-26-6: Inspect/replace fuel and oil hoses (all).

80-6-5: Test Slick impulse couplers.

80-24-3: Inspect/replace the ammeter (all).

81-16-5: Inspect Stick magneto coils.

81-18-4: Replace the oil pump impeller and shaft on Lycoming engines.

82-20-1: Inspect Bendix impulse couplers.

84-26-2: Replace the paper air filter element every 500 hours.

86-17-1: Replace ammeter (supercedes 80-24-3).

Model PA-28 Warrior (151/161)

74-13-4: Install a new end bearing in the throttle control cable (140).

74-14-4: Install weight placard (all through 7415538)

74-24-12: Rework the aileron centering system (1974/75 models).

74-24-13: Replace certain altimeters.

75-8-3: Rework fuel drain valves (all except 235).

75-8-9: Replace the oil pump shaft and impeller on Lycoming engines.

75-16-4: Rework the carburetor air box (1974/75 models)

77-1-1: Check system and placard fuel quantity gauges (all).

77-1-3: Rework the carburetor air box (1974/77).

77-23-3: Check/replace engine and propeller control rod/cable ends.

79-2-5: Inspect/repair the gascolator assembly.

79-7-2: Replace certain batteries.

79-13-3: Inspect fuel line unions.

79-13-8: Replace certain Airborne dry air pumps.

79-18-5: Replace the ELT battery.

79-22-2: Modify the fuel tank vents (all).

79-26-5: Inspect/replace fuel and oil hoses (all).

Model PA-28 Archer II (181)

77-1-1: Check system and placard fuel quantity gauges (all).

77-23-3: Check/replace engine and propeller control rod/cable ends.

79-7-2: Replace certain batteries.

79-13-3: Inspect fuel line unions.

79-13-8: Replace certain Airborne dry air pumps.

79-18-5: Replace the ELT battery.

79-22-2: Modify the fuel tank vents (all).

79-26-5: Inspect/replace fuel and oil hoses (all).

80-6-5: Test Slick impulse couplers.

80-14-2: Repair/replace the throttle linkage.

80-14-3: Modify the PTT switch on the control wheel.

80-24-3: Inspect/replace the ammeter (all).

81-16-5: Inspect Slick magneto coils.

81-18-4: Replace the oil pump impeller and shaft on Lycoming engines.

84-26-2: Replace the paper air filter element every 500 hours.

86-17-1: Replace ammeter (supercedes 80-24-3).

Model PA-28 Dakota (236)

77-12-6: Replace certain Hartzell propeller blades.

79-7-2: Replace certain batteries.

79-13-3: Inspect fuel line unions.

79-13-8: Replace certain Airborne dry air pumps.

79-18-5: Replace the ELT battery.

79-18-6: Comply with Bendix service bulletin.

79-22-2: Modify the fuel tank vents (all).

79-26-5: Inspect/replace fuel and oil hoses (all).

80-14-3: Modify the PTT switch on the control wheel.

80-24-3: Inspect/replace the ammeter (all).

81-16-5: Inspect Slick magneto coils.

81-18-4: Replace the oil pump impeller and shaft on Lycoming engines.

82-11-5: Comply with Bendix service bulletin.

82-20-1: Inspect Bendix impulse couplers.

82-27-3: Every 200 hours, inspect/replace turbine housing if equipped.

84-26-2: Replace the paper air filter element every 500 hours.

86-17-1: Replace ammeter (supercedes 80-24-3).

Model PA-28R Arrow (180/200)

68-12-4: Replace parts in the retraction assembly.

69-12-1: Replace the air induction hose or inspect every 10 hours (all).

69-15-1: Inspect/replace the control wheel pin.

70-9-2: Inspect the propeller spinner for cracks every 25 hours.

70-15-18: Rework autopilot system.

70-26-4: Inspect the stabilator tube with dye penetrant.

71-5-2: Inspect crankshaft bearings, Lycoming IO-360 engines.

71-21-8: Replace the fuel selector cover (all except 235).

72-14-7: Every 100 hours inspect the stabilator attachment fittings.

72-17-5: Modify electric trim control system (if equipped).

73-10-2: Replace Bendix fuel injectors on Lycoming engines.

73-23-1: Inspect piston pins on Lycoming engines.

74-9-4: Move the rear-seat attachment points (all).

74-24-13: Replace certain altimeters.

75-8-3: Rework fuel drain valves (all except 235).

75-8-9: Replace the oil pump shaft and impeller on Lycoming engines.

75-9-15: Install new fuel flow divider gasket on Lycoming engines.

75-24-2: Inspect/replace rear-seat legs and attachments (all).

76-15-8: Install nose gear modification kit PN 761-074V.

77-1-1: Check system and placard fuel quantity gauges (all).

77-12-6: Replace certain Hartzell propeller blades.

77-23-3: Check/replace engine and propeller control rod/cable ends.

79-4-5: Comply with Lycoming service bulletin 433A.

79-7-2: Replace certain batteries.

79-13-8: Replace certain Airborne dry air pumps.

79-18-5: Replace the ELT battery.

79-21-8: Rework lock-nut on injectors.

79-26-5: Inspect/replace fuel and oil hoses (all).

80-19-1: Repair muffler.

80-24-3: Inspect/replace the ammeter (all).

81-11-2: Replace oil drain (all through 1976).

81-18-4: Replace the oil pump impeller and shaft on Lycoming engines.

82-6-11: Rework nose gear (see AD for Serial numbers).

82-13-1: Inspect the Bendix magneto gripper bushings.

82-20-1: Inspect Bendix impulse couplers.

84-26-2: Replace the paper air filter element every 500 hours.

86-17-1: Replace ammeter (supercedes 80-24-3).

Model PA-28R Arrow (201)

77-12-6: Replace certain Hartzell propeller blades.

77-23-3: Check/replace engine and propeller control rod/cable ends.

78-23-10: Replace Bendix fuel-injector seals on Lycoming engines.

79-2-5: Inspect/repair the gascolator assembly.

79-4-5: Comply with Lycoming service bulletin 433A.

79-7-2: Replace certain batteries.

79-13-3: Inspect fuel line unions.

79-13-8: Replace certain Airborne dry air pumps.

79-18-5: Replace the ELT battery.

79-21-8: Rework lock nut on injectors.

79-22-2: Modify the fuel tank vents (all).

79-26-5: Inspect/replace fuel and oil hoses (all).

82-6-11: Rework nose gear (see AD for Serial numbers).

82-13-1: Inspect the Bendix magneto gripper bushings.

82-20-1: Inspect Bendix impulse couplers.

82-27-3: Every 200 hours, inspect/replace turbine housing if equipped.

84-26-2: Replace the paper air filter element every 500 hours.

86-17-1: Replace ammeter (supercedes 80-24-3).

Model PA-32 Cherokee Six (260/300)

66-20-4: Replace the oil filter studs on Lycoming engines.

67-3-7: Replace the fuel selector valve (Ser. 32-151 through 32-535).

67-12-6: Inspect certain tubing for corrosion (all).

67-20-4: Replace the nosewheel torque links.

67-26-2: Rework the fuel lines and selector valves.

67-26-3: Each 200 hours, inspect the fuel system.

67-30-6: Replace the air induction system or inspect every 100 hours.

68-1-3: Applies to seaplane versions only.

69-9-3: Tachometer must be marked with red from 2150 through 2350 rpm to prevent propeller vibration and destruction.

69-15-1: Inspect/replace the control wheel pin.

69-22-2: Inspect the control wheel with dye penetrant.

70-15-18: Rework autopilot system.

70-18-5: Replace landing-gear attach bolts.

70-22-2: Rework the fuel selector valve.

71-9-5: Rework the seat belts.

72-8-6: Inspect the nosewheel torque links (all).

72-14-7: Every 100 hours, inspect the stabilator attachment fittings.

72-17-5: Modify electric trim control system (if equipped).

73-23-1: Inspect piston pins on Lycoming engines.

74-9-2: Comply with Piper Service Bulletin 103, location of the stall warning light.

74-18-13: Nosewheel shimmy/vibration modification.

74-19-1: Inspect the lightening hole in the wing spars (all).

74-24-13: Replace certain altimeters.

75-8-9: Replace the oil pump shaft and impeller on Lycoming engines.

75-9-15: Install new fuel flow divider gasket on Lycoming engines.

75-10-3: Modify the baggage door (1975 models).

75-24-2: Inspect/replace rear-seat legs and attachments (all).

76-18-4: Inspect the fuel selector valve.

77-1-1: Check system and placard fuel quantity gauges (all).

77-12-1: Replace fuel system parts or inspect every 50 hours of operation.

77-12-6: Replace certain Hartzell propeller blades.

77-23-3: Check/replace engine and propeller control rod/cable ends.

78-23-1: Replace fuel drain cover door or inspect every 100 hours.

78-23-10: Replace Bendix fuel-injector seals on Lycoming engines.

79-13-8: Replace certain Airborne dry air pumps.

79-18-5: Replace the ELT battery.

79-21-8: Rework lock-nut on injectors.

79-26-5: Inspect/replace fuel and oil hoses (all).

80-14-1: Modify the fuel vent system to prevent leaks.

80-14-2: Repair/replace the throttle linkage to prevent separation.

80-14-3: Modify the PTT switch on the control wheel.

80-17-14: Comply with Bendix service bulletins.

80-19-1: Repair muffler.

80-24-3: Inspect/replace the ammeter (all).

81-18-4: Replace the oil pump impeller and shaft on Lycoming engines.

82-11-5: Comply with Bendix service bulletin.

82-13-1: Inspect the Bendix magneto gripper bushings.

82-20-1: Inspect Bendix impulse couplers.

82-27-3: Every 200 hours, inspect/replace turbine housing if equipped.

83-22-4: Replace the fuel diaphragm on Bendix fuel injectors on Lycoming engines.

84-26-2: Replace the paper air filter element every 500 hours.

86-17-1: Replace ammeter (supercedes 80-24-3).

Model PA-32-301 Saratoga and Saratoga SP

77-12-6: Replace certain Hartzell propeller blades.

80-14-3: Modify the PTT switch on the control wheel.

80-17-10: Install new wastegate on Lycoming engine.

80-17-14: Comply with Bendix service bulletins.

80-24-3: Inspect/replace the ammeter (all).

81-18-4: Replace the oil-pump impeller and shaft on Lycoming engines.

82-11-5: Comply with Bendix service bulletin.

82-13-1: Inspect the Bendix magneto gripper bushings.

82-20-1: Inspect Bendix impulse couplers.

83-22-4: Replace the fuel diaphragm on Bendix fuel injectors on Lycoming engines.

84-26-2: Replace the paper air filter element every 500 hours.

86-17-01: Replace ammeter (supercedes 80-24-3).

Model PA-32R-300 Lance

76-11-9: Reroute the fuel lines located in the wheel wells to prevent chafing.

76-15-8: Install nose-gear modification kit PN 761-074V.

76-16-08: Replace the engine oil coolers to prevent oil starvation.

76-18-4: Inspect/replace the fuel selector valve.

77-12-6: Replace certain Hartzell propeller blades.

77-23-3: Check/replace engine and propeller control rod/cable ends.

78-16-8: Replace oil cooler.

78-22-7: Replace the control-column stop sleeve.

78-23-1: Replace fuel drain cover door or inspect every 100 hours.

78-23-10: Replace Bendix fuel-injector seals on Lycoming engines.

79-13-4: Install new fuel-flow indicator tubes (Ser. 7887001 through 7987124).

79-13-8: Replace certain Airborne dry air pumps.

79-18-6: Comply with Bendix service bulletins.

79-26-4: Rework/modify the rudder to prevent skin cracks.

79-26-5: Inspect/replace fuel and oil hoses (all).

80-14-1: Replace leaking fuel-tank vents.

80-14-2: Modify the PTT switch on the control wheel.

80-17-10: Install new wastegate on Lycoming engine.

80-17-14: Comply with Bendix service bulletins.

80-20-5 Inspect/replace the turbo exhaust coupling by penetrant inspection.

80-24-3: Inspect/replace the ammeter (all).

80-24-7: Modify the gear-down locks.

81-18-4: Replace the oil-pump impeller and shaft on Lycoming engines.

81-19-4: Inspect/replace all hoses.

81-24-7: Modify the nosewheel.

82-11-5: Comply with Bendix service bulletin.

82-20-1: Inspect Bendix impulse couplers.

83-22-4: Replace the fuel diaphragm on Bendix fuel injectors on Lycoming engines.

84-26-2: Replace the paper air filter element every 500 hours.

86-17-1: Replace ammeter (supercedes 80-24-3).

Model PA-24 Comanche (180/250/260/400)

58-25-5: Modify the door latch (all through 24-336).

59-6-5: Replace the nose-gear bungees (all through 24-503).

59-7-5: Rework the oil cooler lines on 180 models.

59-12-9: Inspect/replace the control wheel sprocket (all through 24-885).

59-13-2: Reinforce the aileron balance-weight attachment (all through 24-980).

59-26-2: Modify the fuel-cell vent tube (all through 24-1373).

60-24-3 Modify the fuel-cell vent tube (all through 24-2161).

61-16-6: Replace the fuel-tank selector valve on all models with wing tanks.

61-20-2: Reinforce the exhaust stack (ser. 24-103 through 1629) 250-hp only.

62-10-3 Remove the "Rubatex" blocks from the travel area of the ailerons (all through 24-2264).

62-26-5: Rework complete exhaust system (all through 24-3284).

63-22-3: Rework all with Marvel carburetor.

63-27-3: Replace the landing gear circuit breaker.

64-10-4: Modify the carburetor air box (Ser. 24-1477 through 24-3646).

64-16-5: Modify the fuel pump on Lycoming engines.

65-2-3: Replace turbine parts or inspect every 100 hours.

65-11-4: Rework stabilator control system if equipped with autopilot.

66-18-4: Rework the baggage compartment door latch.

66-20-4: Replace the oil filter studs on Lycoming engines.

67-22-6: Replace the fuel-injector diaphragm in Bendix fuel-injection systems on Lycoming engines.

68-5-1: Inspect the exhaust system every 50 hours.

68-13-3: Install modification kit or inspect fuel system every 40 hours.

68-19-4: Service propeller blades every 1000 hours.

70-15-18: Rework autopilot system.

71-12-5: Modify the electric trim control.

72-22-5: Placard maximum speed (all), or install balance weights to tail control surfaces.

73-2-1: Inspect/rework the propeller blades every 1000 hours.

73-10-2: Replace Bendix fuel injectors on Lycoming engines.

73-23-1: Inspect piston pins on Lycoming engines.

74-13-1: Inspect/repair the stabilator torque-tube supports.

74-13-3: Every 500 hrs/3 years, inspect the stabilator attach-bolts for corrosion.

74-24-13: Replace certain altimeters.

75-5-2: Modify the Beryl oil filter.

75-8-9: Replace the oil-pump shaft and impeller on Lycoming engines.

75-9-15: Install new fuel-flow divider gasket on Lycoming engines.

75-12-6: Inspect the fin-spar attachment each 100 hours of operation.

75-27-8: Install modification kit on stabilator torque-tube bearing blocks.

76-19-7: Replace the stabilator weight assembly.

77-8-1: Inspect the aileron hinge brackets every 100 hours.

77-9-10: Modify the electric trim control.

77-12-6: Replace certain Hartzell propeller blades.

77-13-21: Replace landing-gear bungees every 500 hours and inspect entire landing gear every 1000 hours.

77-16-1: Disassemble and inspect propeller IAW AD.

79-4-5: Comply with Lycoming Service Bulletin 433A.

79-12-8: Inspect the fuel selector for water contamination every 50 hours.

79-13-8: Replace certain Airborne dry air pumps.

79-18-5: Replace the ELT battery.

79-20-10: Replace aileron nose-rib or inspect every 100 hours.

79-21-8: Rework lock-nut on injectors.

80-6-5: Test Slick impulse couplers.

81-15-3: Replace the Brackett engine air filter.

81-18-4: Replace the oil-pump impeller and shaft on Lycoming engines.

82-19-1: Inspect the wing lower-main-spar caps and upper-main-spar attachment every 100 hours.

81-19-4: Inspect/replace all hoses.

82-19-1: Inspect the wing spar caps and plates with dye penetrant.

82-20-1: Inspect Bendix impulse couplers.

82-23-1: If Robertson STOL, modify flap system.

83-10-1: Modify the fuel system to prevent contamination (400).

83-19-3: Inspect the spar cap.

84-26-2: Replace the paper air filter element every 500 hours.

Model PA-38 Tomahawk

78-22-1: Install rudder kit 763-872 to allow for increased surface clearance.

79-23-4: Modify the wing spar attach-fittings per service bulletin by replacing missing rivets.

78-23-9: Modify or replace the control wheels (all through Ser. 38-78A0329) to prevent jamming.

78-25-1: Replace Slick magnetos.

78-26-6: Replace the vertical-fin spar plate.

79-3-2: Install rudder hinge kit (all through Ser. 80A0063).

79-8-2: Inspect stabilizer fittings and torque bolts (all through Ser. 79A0312) with penetrant.

79-13-8: Replace certain Airborne dry air pumps.

79-17-5: Rewire the instrument panel.

79-18-5: Replace the ELT battery.

80-6-5: Test Slick impulse couplers.

80-22-13: Replace the rudder hinge brackets (all through Ser. 80A0165).

80-25-2: Inspect valve pushrods on Lycoming 0-235 engines.

80-25-7: Replace certain oil coolers.

81-4-7: Inspect/repair the forward fin-spar attachment web fitting.

81-16-5: Inspect the Slick magneto coils.

81-18-4: Replace the oil-pump impeller and shaft on Lycoming engines.

81-23-7: Replace the engine mount (all through Ser. 78A0678).

82-2-1: Install aileron-balance kit (all through Ser. 81A0051).

82-27-8: Replace the fin spar (all through Ser. 82A0122).

83-5-4: Replace the landing-gear bolts (all through Ser. 82A0110).

83-14-8: Replace the airspeed indicator.

83-19-1: Modify fin spar IAW service bulletin 763A or inspect every 100 hours.

84-26-2: Replace the paper air filter element every 500 hours.

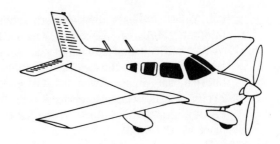

Chapter 9
STCs and Modifications

No matter what the product, be it airplane, boat, or automobile, someone wants to make a modification or change to it in order to improve upon the basic design, or to add a feature not factory-available. Piper airplanes are no different.

As all production airplanes are Type Certificated by the FAA, it is necessary to obtain authorization from the same authority to make changes to the basic aircraft. The vehicle for these changes is the Supplemental Type Certificate, or STC.

THE SUPPLEMENTAL TYPE CERTIFICATE

The STC is the authorization to change the original aircraft to a modified configuration. Examples of these changes would be the addition of an autopilot, installation of a larger engine, or even a wing modification for better low-speed performance (STOL—short take-off and landing).

The following is a list of all currently approved STCs for the PA-24, PA-28, PA-32, and PA-38 airplanes.

PA-24 Series

SA74SW: Quick-door-release mechanism (all PA-24); Ross Aviation, Inc., R.F.D. 5, Riverside Airport, Tulsa, OK 74101.
SA341CE: SAGA (semi-automatic gear actuator) control system

(all PA-24); In-Flight Devices Corporation, Port Columbus International Airport, Columbus, OH 43219.

SA478EA: SAGA (semi-automatic gear actuator) system (all PA-24); In-Flight Devices Corporation, Port Columbus International Airport, Columbus, OH 43219.

SA482WE: Plastic wingtips replaced with "Speedplates" (all PA-24); Grand Valley Aircraft, Inc., P.O. Box 158, Grand Junction, CO 81501.

SA526GL: Installation of aileron, flap, fuselage, and rudder gap seals, wing root fairings, and dorsal fin (all PA-24); Knots 2 U, Inc., 1941 Highland Avenue, Wilmette, IL 60091.

SA871WE: Fiberglass wingtips (all PA-24); Met-Co-Aire, P.O. Box 2216, Fullerton, CA 92633.

SA891WE: Toe brakes (all PA-24); Atlas Aviation, Inc., 8505 Montview Blvd., Denver, CO 80202.

SA208GL: Installation of one-piece windshield (PA-24-180, -250, -260, -400): Great Lakes Aero Products, 2412 Davison Rd., Flint, MI 48506.

SA270GL: Installation of forward windows, P/Ns W/T-, WG-2058-187 or -250 and W-, W/G20599-187 or -250 (PA-24-250, S/N 24-2299 and above; PA-24-260, S/N 24-3642, 24-4000 through 24-4802; PA-24-400, S/N 26-2 and above); Great Lakes Aero Products, 2412 Davison Rd., Flint, MI 48506.

SA891SO: Remove existing windshield and install single-piece windshield (PA-24, all models with two-piece windshield): Beryl D'Shannon Aviation Specialties, Inc., P.O. Box 76155, Atlanta, GA 30328.

SA2272NM: Installation of fiberglass window mouldings (PA-24-250, -260, S/N 24-1 through 24-4299; PA-24-400); Serv-Aero Engineering, Inc., 37 Mortenser Avenue, Salinas, CA 93905.

SA2541NM: Installation of fiberglass wingtips (PA-24-250, -260, -400); Johnston Aircraft Services, Inc., P.O. Box 1557, Tulare, CA 93275.

SA3070WE: Installation of single-piece windshield (PA-24-250, -260, -400); Wayne Airframe, Inc., 1644 Roscoe Boulevard, Van Nuys, CA 91406.

SA2495WE: Installation of recontoured wing leading edge, stall fences, drooped ailerons in flaps-down mode, flap-actuated stabilator trim system, optional new wingtips, optional SC-100 stall warning system (PA-24-250, -400); Robertson Aircraft Corp., 15400 Sunset Highway, Bellevue, WA 98007.

SA3565WE: Installation of single-piece windshield (PA-24-250, -260, -400); Cee Bailey's Aircraft Plastics, 2955 Junipero Avenue, Long Beach, CA 90806.

SA4179WE: Installation of inflatable door seal (PA-24-250, -260, -400); Bob Fields Aerocessories, 340 "E" East Santa Maria Street, Santa Paula, CA 93060.

SA1152SW: One-piece windshield (PA-24-250, -260, -400); J.W. Miller Aviation, Inc., HSB Box 7757, Marble Falls, TX 78654.

SA1153SW: Square wingtips with dual navigation lights (PA-24-250, -260, -400); J.W. Miller Aviation, Inc., HSB Box 7757, Marble Falls, TX 78654.

SA31NW: Installation of Madras wingtips (PA-24-260, -400); Madras Air Service, Route 2 Box 1225, Madras, OR 97741.

SA1-340: Oil-free vacuum pump PA-201 (all PA-24); Associated Aircraft Tool Manufacturing, Inc., 3660 Dixie Highway, Hamilton, OH 45012.

SA4-852: Full-flow oil filter 30409A with element 1A0325 (all PA-24); Winslow Aerofilter Corp., 4069 Hollis Street, Oakland, CA 94608.

SA4-1032: Aerofilter 30409A full-flow lube oil filter with element 1A0235 (all PA-24); Winslow Aerofilter Corp., 4069 Hollis Street, Oakland, CA 94608.

SA510CE: Modify electrical system to combination 12/24 volt system for starting engine (all PA-24); AVCO Lycoming Division, AVCO Corp., Williamsport, PA 17701.

SA511WE: Lycoming O-540-B1B5 engine, Hartzell HC-A2XK-1/8433-7 propeller required (all PA-24); Johnston Aircraft Service, P.O. Box 757, Tulare, CA 93274.

SA904WE: Hanlon and Wilson exhaust system, 637-129 (all PA-24); "Slim" Kidwell Aviation Company, Municipal Airport, Torrance, CA 90510.

SA928NW: Aeroquip high-temperature hydraulic hoses and fittings between oil cooler and engine (all PA-24); Don Rudebaugh, 409 North East 12th Avenue, Vancouver, WA 98664.

SA1109WE: Hanlon and Wilson Exhaust system (all PA-24); "Slim" Kidwell Aviation Company, Municipal Airport, Torrance, CA 90510.

SA4582SW: Installation of JASC6550 12-volt 50-amp alternator,

J12-100 SP voltage regulator, and associated wiring (PA-24-180); Skytronics, Inc., 227 Oregon Street, El Segundo, CA 90245.

SA2653WE: Installation of engine crankcase, breather, and vacuum pump air-oil separator (PA-24-250, -260); Beryl D'Shannon Aviation Specialties, Inc., P.O. 76155, Atlanta, GA 30328.

SA3376WE: Installation of flexible stainless steel oil cooler hoses (PA-24-250, -260); Aircraft Metal Products Corp., 4206 Glencoe Ave., Venice, CA 90291.

SA5-27: Remove Lycoming O-540-A1A5 (250-hp) engine and install Lycoming O-540-B1B5 (235-hp) engine (PA-24-250); Peninsula Airways, Inc., P.O. Box 110, Nak Nak, AK 99633.

SA811WE: Installation of turbosupercharged Lycoming IO-540-C1B5 or C1C5 engine (PA-24-250); Roto-Masters, Inc., 7101 Fair Avenue, North Hollywood, CA 91605.

SA3684WE: Installation of an air/oil separator (PA-24-250, -260); Walker Engineering Company, 2240 Sawtelle Blvd., Los Angeles, CA 90064.

SA893SO: Installation of exhaust silencers (PA-24-260); Beryl D'Shannon Aviation Specialties, Inc., P.O. Box 76155, Atlanta, GA 30328.

SA2062WE: Installation of Lycoming IO-540-R1A5 engine and Rajay Turbo-superchargers (PA-24-260); Rajay Industries, Inc., 2602 East Wardlow Rd, Long Beach, CA 90807.

SA2359WE: Installation of turbosupercharged Lycoming IO-720-A1A engine (PA-24-400); Roto-Master, Inc., 7101 Fair Ave., North Hollywood, CA 91605.

SA4-666: Grimes D-7080-1-12 rotating beacon (all PA-24); H.M. Ruberg, 1300 North 28th St., Springfield, OR 97477.

SA4-955: Fiberglass vertical-fin mount for either Grimes or Whelen WRM rotating beacons (all PA-24); Art Whitaker, Pearson Airpark, Vancouver, WA 98660.

SA4-956: Three seat belts to accommodate three persons in the rear seat (all PA-24); Bates Aircraft, Inc., Municipal Airport, Hawthorne, CA 90250.

SA2SO: Fifth passenger seat (all PA-24); Red Aircraft Service, Inc., International Airport, Ft. Lauderdale, FL 33300.

SA48WE: Belly-mounted rotating anti-collision light to supplement top-fuselage-mounted standard rotating light (all PA-24); Stits Aircraft Co., P.O. Box 3084, Riverside, CA 95202.

SA112CE: APU receptacle (all PA-24); John O. Grimm, 3648 Eminence Blvd., St. Louis County, MO 63155.

SA113CE: APU receptacle (all PA-24); John O. Grimm, 3648 Eminence Blvd., St. Louis County, MO 63155.

SA203SO: Fyr-Fyter 23-4, 2½ pound dry chemical fire extinguisher (all PA-24); Memphis Aero Corp., Municipal Airport, Memphis, TN 38130.

SA315SW: Fabrication and installation of split rear-seat back (all PA-24); Seven Bar Flying Service, Inc., 10001 Coors Rd. NW, Albuquerque, NM 87101.

SA881EA: Installation of Grimes 30-0437-1 anti-collision strobe light kit, P/N 35-0200-1 (all PA-24); Grime Manufacturing Co., 515 North Russell St., Urbana, OH 43078.

SA108NE: Replacement of brake discs, Piper P/N 751672 (PA-24-250); William C. Roberts, 5 Lynbrook La, Doylestown, PA 18901.

SA1315CE: Chrome-plated brake disc installations (PA-24-250, -260); Engineering Plating and Processing, Inc., 641 Southwest Blvd., Kansas City, KS 66103.

SA1745SO: Installation of stainless steel brake discs (PA-28 series); Appalachian Accessories, P.O. Box 1077, Tri-City Airport Station, Blountville, TN 37617.

SA1706SW: Portable solid oxygen system (PA-24-250), -260, -400); Stand-By Systems, Inc., 4907 Brookview Dr., Dallas, TX 75220.

SA1-554: External electrical power receptacle (PA-24-250); R.H. Haynes, Inc., 62 Boorhis La, Hackensack, NJ 07601.

SA1115SW: Dual brakes (PA-24-260, -400); J.W. Miller Aviation, Inc., HSB Box 7757, Marble Falls, TX 78654.

SA331SO: External power supply plug (PA-24); Montgomery Aviation Corp, P.O. Box 92, Montgomery, AL 36100.

SA1689SO: Installation of shoulder harnesses (PA-24); Kosola and Associates, Inc., 1302 West Broad Ave., Albany, GA 31706.

SA4-971: 20-gallon auxiliary fuel tank in baggage compartment (all PA-24); Symons Engineering, P.O. 90002, Airport Station, Los Angeles, CA 90009.

SA4-1235: Installation of Brittan 15-gallon wingtip fuel tanks (all PA-24); H.S. Osborne, d/b/a Osborne Tank and Supply, Star Route, Box 12, Oro Grande, CA 92368.

SA4-1248: Fuel drain modified to incorporate a handle extending forward through front bulkhead (all PA-24); Bates Aviation, Inc., 1026 East 120th Street, Hawthorne, CA 90252.

SA41351: Installation of wingtip fuel tanks (all PA-24); H.S. Osborne, d/b/a Osborne Tank and Supply, Star Route, Box 12, Oro Grande, CA 92368.

SA18SO: Cockpit-operated fuel drain valve (all PA-24); Red Aircraft Service, Inc., International Airport, Ft. Lauderdale, FL 33300.

SA4-3151: Wingtip fuel tanks installed (PA-24-180); Brittain Industries, Inc., 2700 Skypark Dr., Torrance, CA 90505.

SA1178EA: Installation of Silver Instruments Fueltron IG or IP or Fuelguard digital fuel-flow and totalizer system (PA-24-250, -260); Silver Instruments, Inc., 2346 Stanwell Dr., Concord, CA 94520.

SA3773WE: Installation of SDI model CFS-1000, -1001, FT-100, or FT-101 fuel-flow indication system (fuel-injected engines only) (PA-24-250, -260); Silver Instruments, Inc., 1762 McGaw Ave., Irvine, CA 92714.

SA5758SW: Install ALCOR fuel-flow/totalizer system 24606 which includes transducer 86216 24606, indicator, and related hardware (PA-24-250, -260); ALCOR, Inc., 10130 Jones-Maltsberger, P.O. Box 32516, San Antonio, TX 78284.

SA4361WE: Installation of Silver Instruments Fueltron IG or Fueltron IP digital fuel-flow indicating system (PA-24-400); Silver Instruments, Inc., 2346 Stanwell Dr., Concord, CA 94520.

SA598EA: Installation of electronic voltage alarm, Drawing Number 80107 (all PA-24); McKinnley Engineering, Corp, P.O. 275, Palisades Park, NJ 07650.

PA-28 Series

SA8PC: Installation of baggage compartment kit IFC-BC-001 behind rear seats in the cabin (PA-28 series); Island Flight Center, Inc., 203 Lagoon Dr., Honolulu, HI 96819.

SA224CE: A3000 main and NB1200 nose ski (PA-28 series); Fluidyne Engineering Corp, 5900 Olson Memorial Hwy, Minneapolis, MN 55422.

SA246SW: Nose cowling modified (PA-28 series); Hutchinson Aircraft Service, P.O. Box 524, Borger, TX 79006.

SA319CE: Model 2500 wheel skis (PA-28 series); Fluidyne

Engineering Corp, 5900 Olson Memorial Hwy, Minneapolis, MN 55422.

SA481SW: Removal of door for parachute jumping (PA-28 series); Irvin J. Fusilier, 9617 Fulton St., Houston, TX 77002.

SA1191WE: Removal of baggage door for aerial photography operations (PA-28 series); Kelsey Ellis Air Service, Inc., Municipal Airport, Salt Lake City, UT 84101.

SA2325WE: Replacement of Station 140.6 vinyl bulkhead panel with fiberglass baggage compartment (PA-28-140); Chaffee Aircraft Service Corp., 2105 Valley Blvd., Colton, CA 92324.

SA2NW: Installation of Van's Aircraft Model VA-1 fiberglass wheel fairings (PA-28-140, -150, -160, -180, -235); Van's Aircraft, Route 2 Box 187, Forrestt Grove, OR 97116.

SA549NW: Installation of Deemers wingtips (PA-28-140, -150, -160, -180); Madras Air Service, Route 2 Box 1225, Madras, OR 97741.

SA603GL: Installation of aileron and flap gap seals (PA-28-140, -150, -160, -180, -235; PA-28R-180, -200); Knots 2 U, Inc., 1941 Highland Ave., Wilmette, IL 60091.

SA708SW: Plane Booster safe flight wingtips 101-28 (PA-28-140, -150, -160, -180); Harvey J. Ferguson, d/b/a Plane Booster, Inc., P.O. Box 564, McAllen, TX 78501.

SA780GL: Installation of aileron and flap gap seals (PA-28-140, -150, -160, -180; PA-28R-180, -200); General Aviation Corp., Rock County Airport, Janesville, WI 53545.

SA880WE: Installation of a new design fiberglass wingtip (PA-28-140, -150, -160, -180; PA-28R-180, -200); Met-Co-Air, P.O. Box 2216, Fullerton, CA.

SA1072CE: Installation of wing leading edge cuffs and droop tips, dorsal fin and vertical stabilizer vortex generator (PA-28-140, -150, -160, -180; PA-28R-180, -200); Horton STOL-Craft, Inc., Wellington Municipal Airport, Wellington, KS 67152.

SA1227CE: Installation of Dorsal fin (PA-28-140, -150, -160, -180); Isham Aircraft, P.O. Box 12172, Wichita, KS 67212.

SA1228CE: Installation of wingtip extension (PA-28-140, -150, -160, -180); Isham Aircraft, 416 West 4th St., Valley Center, KS 67147.

SA1373CE: Install third window (PA-28-140, -150, -160, -180, -235); Isham Aircraft, P.O. Box 12172, Mid-Continent Airport, Wichita, KS 67277.

SA1487SO: Installation of speed enhancement kit (PA-28-140, -150, -151, -160, -161, -180, -181, -235, -236; PA-28R-180,

-200); Sea Wings, Inc., 1-1 Wintberg Skyline Dr., St. Thomas, VI.

SA1696SO: Installation of one-piece windshield (PA-28-140, -150, -151, -160, -161, -180, -181, -235, -236, -201T; PA-28R-180, -200, -201, -201T; PA-28RT-201, -201T); Kosola and Associates, Inc., 1302 West Broad Ave., P.O. Box 3452, Albany, GA 31706.

SA1741WE: Installation of hand control for rudder system (PA-28-140, -150, -160, -180, -235; PA-28R-180, -200); William Henry Blackwood, Route 3 Box 744, B-3, Escondido, CA 92025.

SA1898SW: Installation of wing leading edge cuff, flow fences, optional wingtips and dorsal fin (PA-28-140, -150, -160, -180, -235; PA-28R-180, -200); Barbara or Bob Williams, Box 431, 213 North Clark, Udall, KS 67146.

SA2179WE: Installation of leading-edge cuff, droop ailerons, wing stall fences, raked wingtips, fuselage flaps, stabilator trim spring, and dorsal fin (optional) (PA-28-140, -150, -160, -180); Charyl C. Robertson, 15400 Sunset Hwy., Bellevue, WA 98007.

SA4236WE: Installation of inflatable door seal (PA-28-140, -150, -151, -160, -161, -180, -181, -236, -201T; PA-28R-180, -200, -201, -201T; PA-28RT-201); Bob Fields Aerocessories, 5673 Stanford St., Ventura, CA 92003.

SA640GL: Installation of aileron and flap gap seals (PA-28-151, -161, -181, -236; PA-28R-201, -201T; PA-28RT-201, -201T); Knots 2 U, Inc., 1941 Highland Ave., Wilmette, IL 60091.

SA855GL: Installation of aileron and flap gap seals (PA-28-151, -161, -181, -236; PA-28R-201, -201T; PA-28RT-201, -201T); General Aviation Corp., Rock County Airport, Janesville, WI 53545.

SA1075SO: ''Wing-Ding'' door stop 2B installed using existing screws (Piper P/N 415 309, MS27039-0809) on leading edge on right wing at inboard side of fuel cell (W.S.5700) (PA-28-151, -161, -181, -235; PA-28R-200, -201, -201T); William J. Stephenson, Pro-Flite of Vero Beach, Inc., P.O. Box 998, Vero Beach, FL 32960.

SA1463SO: Installation of Speed Enhancement Kit (PA-28-161); Causey Aviation Service, Inc., Route 1 Box 137, Liberty, NC 27298.

SA1607SO: Fabrication and installation of nosewheel fairings

(PA-28-161, -181); Windy's Aircraft Parts, 3508 Greenview Ave., Rainbow City, AL 35901.

SA2202WE: Installation of contoured leading edge, droop ailerons, fuselage flap, stabilator trim spring, wing stall fences, raked wingtips, and dorsal fin (optional) (PA-28R-180, -200); Charyl C. Robertson, 15400 Sunset Hwy., Bellevue, WA 98007.

SA2490SW: Wingtip extension and dorsal fin (PA-28R-180, -200); Isham Aircraft, P.O. Box 12172, Mid-Continent Airport, Wichita, KS 67277.

SA1580SO: Installation of speed enhancement kit (PA-28R-201; PA-28RT-201, -201T); Sea Wings, Inc., 1-1 Wintberg Skyline Dr., St. Thomas, VI.

SA2171NM: Installation of Precise Flight Speedbrake System (PA-28-201T, PA-28R-201T, PA-28RT-201T); Precise Flight, Inc., 63120 Powell Butte Rd., Bend, OR 97701.

SA1580SO: Installation of speed enhancement kit (PA-28RT-201T); Sea Wings, Inc., 1-1 Wintberg Skyline Dr., St. Thomas, VI.

SA2143WE: Installation of drooped leading edge, drooped aileron, wing stall fences, fuselage flap, droop wingtips, and elevator trim spring (PA-28-235); Robertson Aircraft Corp., 15400 Sunset Hwy., Bellevue, WA 98004.

SA281AL: Installation of large nose-gear fork and 8.00-6 nose-gear tire (PA-28 series); Tibbetts-Agree Airmotive, P.O. Box 110, Nak Nak, AK 99633.

SA1-468: Six- or 13-quart propeller anti-icing kit WAP-101A (PA-28 series); E.W. Wiggins Airways, Inc., Municipal Airport, Norwood, MA 02062.

SA1-649: Lube oil filter PB55-1 (PA-28 series); Fram Corp., 105 Pawtucket Ave., Providence, RI 02916.

SA222SW: Hartzell HC-82XL-6N/7636D-4 propeller and cowl rework (PA-28 series); Hutchinson Aircraft Service, P.O. Box 1070, Borger, TX 79006.

SA556SW: Constant-speed propeller (PA-28 series); Hutchinson Aircraft Service, P.O. Box 1070, Borger, TX 79006.

SA913EA: Installation of Elano Corp. heater-muffler system (PA-28-140); Elano Corp., 2455 Dayton-Xenia Rd., Xenia, OH 45385.

SA2052WE: Installation of Hartzell HC82XL-6F/8433-12 propeller (after conversion of engine to Lycoming O-320-E1A model) (PA-28-140); Propellers, Inc., 5802 South 228th St., Kent, WA 98031.

SA2706SW: Lycoming O-320-D3G engine (PA-28-140); RAM Aircraft Modifications, P.O. Box 5219, Waco, TX 76708.

SA3196WE: Installation of Lycoming O-320-D1A engine, Hartzell HC-C2YL-1BF/F7663-4 or Hartzell HC-C2YL-1BF/8468A-8R propeller and associated powerplant components (PA-28-140); John Grodahl, 4224 W. Ash, Fullerton, CA 92633.

SA3415WE: Installation of Lycoming 0-320-D2A engine and Sensenich M74DM6-0-60 propeller (PA-28-140); Arthur M. D'Onofrio, Jr., 206 Winthrop Blvd., Cromwell, CT 06416.

SA1331CE: Installation of 160-hp Lycoming engine and repitched Sensenich propeller (PA-28-140, -150, -151); Schneck Aviation, Inc., Greater Rockford Airport, P.O. Box 6417, Rockford, IL 61125.

SA1963CE: Operation on unleaded and/or leaded automotive gasoline (PA-28-140, -150, -151); Petersen Aviation, Inc., Route 1 Box 18, Minder, NE 68959.

SA793CE: Installation of Lycoming O-360-A1A engine and Hartzell HC-C2YK-1B/7666A-2 or HC-C2YK-1BF/F7666A-2 propeller (PA-28-140, -150, -151, -160, -161); Robert L. And Barbara V. Williams, Box 654, Udall, KS 67146.

SA802GL: Modify airplane to fly on unleaded automotive gasoline, 87 minimum antiknock index (PA-28-140, -150, -151); Petersen Aviation, Inc., Route 1 Box 18, Minder, NE 68959.

SA3435WE: Installation of flexible oil hose assembly (PA-28 series); Aircraft Metal Products Corp., 4206 Glencoe Ave., Venice, CA 90291.

SA3687WE: Installation of an air/oil separator (PA-28 series); Walker Engineering Co., 2240 Sawtelle Blvd., Los Angeles, CA 90064.

SA4244WE: Installation of Elano P/NN EL099001-072 "NC" (or later FAA approved revision) muffler in lieu of original Avcon muffler (PA-28-140, -150, -151, -160); Del Air, P.O. Box, Strathmore, CA 93267.

SA228GL: Installation of Lycoming O-320-D3G engine in lieu of Lycoming O-320-E3D engine using original installation hardware and components. No installation data required (PA-28-151); W. Irvine Young, 54 Atomic Ave., Toronto, Ont, Canada M8Z5L4.

SA2969SW: Lycoming O-320-D3G engine (PA-28-151); RAM Aircraft Modifications, P.O. Box 5219, Waco, TX 76708.

SA180EA: Geared starter (PA-28-160); Turner Field, Inc., Prospectville, PA 19077.

SA2213WE: Conversion of Lycoming O-360-A3A engine to Model O-360-A1A engine and installation of Hartzell HC-C2YK1-B/7666A-0 propeller (PA-28-180); Propellers, Inc., 5802 South 228th St., Kent, WA 98031.

SA194GL: Installation of Lycoming IO-360-C1C6 engine in lieu of Lycoming IO-360-C1C engine using original installation hardware and components (PA-28R-200); Shelby Aircraft Engine Parts, Inc., P.O. Box 454, Shelbyville, IL 62565.

SA5679SW: Electrically driven vacuum pump as standby auxiliary pump (PA-28R-200); Aero Safe Corp., Box 10206, Fort Worth, TX 76114.

SA1762NM: Installation of a pressurized magneto system on the Continental turbocharged TSIO-360-C, -F, and -FB engines (PA-28-201T, PA-28R-201T, PA-28RT-201T); S.G.H. Inc., 1737 West Valley Hwy., Auburn, WA 98002.

SA2145NM: Installation of Turboplus intercooler system PA-28-9000 (PA-28-201T; PA-28R-201T; PA-28RT-201T); Turboplus, Inc., 1437 West Valley Hwy., Auburn, WA 98002.

SA2148NM: Installation of Turboplus engine nacelle cowl flaps (PA-28-201T; PA-28R-201T; PA-28RT-201T); Turboplus, Inc., 1437 West Valley Hwy., Auburn, WA 98002.

SA5681SW: Installation of electrically driven vacuum pump as a standby auxiliary pump to the existing instrument air system (PA-28-201T; PA-28R-201T; PA-28RT-201T); Aero Safe Corp., P.O. Box 10206, Fort Worth, TX 76114.

SA1383CE: Installation of Edo-Aire propeller governor Model 34-828-014-12 (PA-28R-201T); Edo-Aire Wichita Division, 1326 South Walnut St., Wichita, KS 67213.

SA2144NM: Installation of a modified air-induction system to the Continental TSIO-201T engine turbocharger compressor (Rajay) system (PA-28R-201T; PA-28RT-201T); Turboplus, Inc., 1437 West Valley Hwy., Auburn, WA 98002.

SA2147NM: Installation of Continental TSIO-360-FB(C) (converted) engine and associated system (PA-28-201T, PA-28R-201T, PA-28RT-201T); Turboplus, Inc., 1437 West Valley Hwy., Auburn, WA 98002.

SA2167NM: Installation of the Precise Flight standby vacuum system (SVS) (PA-28R-201T); Precise Flight, Inc., 63120 Powell Butte, Rd., Bend, OR 97701.

SA5681SW: Electrically driven vacuum pump as standby auxiliary pump (PA-28RT-201T); Aero Safe Corp., Box 10206, Fort Worth, TX 76114.

SA1964CE: Operation on unleaded and/or leaded automotive gasoline (PA-28-235); Petersen Aviation, Inc., Route 1 Box 18, Minden, NE 68959.

SA1189SW: Air circulator (PA-28-140); Ves-Kol. 2805 National Dr., Garland, TX 75040.

SA2285NM: Installation of Cessna control wheels, P/N 0513260-4 and associated installation components (PA-28-140); John H. Lund, 3833 West Harmont, Phoenix, AZ 85021.

SA2842WE: Installation of Novastar anti-collision lighting system (PA-28-140); Symbolic Displays, Inc., 1762 McGaw Ave., Irvine, CA 92705.

SA600NW: Installation of forward-facing white lights, clear plastic wingtip-tank nose cone. (*Note:* Installation does not apply to those models with tapered wings.) (PA-28-140, -150, -160, -180, -235); Robert C. Cansdale, Bob's Aircraft Supply, Thun field, Puyallup, WA 98371.

SA819EA: Alteration to replace standard anti-collision light with Grimes Series 550 single anti-collision red or white strobe kit, P/N 34-0001-1 or 34-0001-2 (PA-28-140, -150, -160; PA-28R-180, -200); Grimed Manufacturing Co., 515 North Russell St., Mineral Wells, TX 76067.

SA1123EA: Installation of DeVore vertical-tail floodlights on upper surface of horizontal stabilator (PA-28 series); DeVore Aviation Corp., Suite B, 6104 Kircher St., NE, Albuquerque, NM 87109.

SA1317CE: Chrome-plated brake disc installation (PA-28-140, -150, -160, -180, -235; PA-28R-180, -200); Engineering Plating and Processing, Inc., 641 Southwest Blvd, Kansas City, KS 66103.

SA3233WE: Installation of Symbolic Displays "Novastar" anti-collision light (PA-28-180); Symbolic Displays, Inc., 1762 McGaw Ave., Irvine, CA 92705.

SA1258EA: Installation of Cosco Model 78 child restraint system (PA-28R-200); Stuart R. Miller, P.O. Box 926, Grand Central Station, New York, NY 10163.

SA472SO: Installation of hand control for rudder pedal operation (PA-28 series); The Wheelchair Pilots Association, 4211 Fourth Ave. S., St. Petersburg, FL 33711.

SA178WE: Full-flow aerofilter 30409A with element 1A0235

(PA-28 series); Winslow Aerofilter Corp., 4069 Hollis St., Oakland, CA 94608.

SA3071WE: Installation of Bendix P/N 480543 electric auxiliary fuel pump (PA-28-140); Harry R. Dellicker, P.O. Box 746, Strathmore, CA 93267.

SA2223NM: Installation of SDI Model CFS-1000A, -1001A, FT-100, or FT-101 fuel-flow indicating system (PA-28-140, -150, -151, -160, -161); Symbolic Displays, Inc., 1762 McGaw Ave., Irvine, CA 92714.

SA2280NM: Installation of Silver Instruments Fueltron IG-CS, IP-CS, IL-CS, or Fuelgard fuel-flow indicating system (PA-28-140, -150, -151, -160, -161); Silver Instruments, Inc., 8202 Capwell Dr., Oakland Airport Business Park, Oakland, CA 94621.

SA4276WE: Installation of SDI CFS-1000, -1001, or FT-100 fuel-flow indicating system and P/N 480543 auxiliary fuel pump (PA-28-140, -150, -180); Del Air, P.O. Box 746, Strathmore, CA 93267.

SA3841WE: Installation of SDI Model CFS-1000A, -1001A, FT-100, or FT-101 fuel-flow indicating system (PA-28R-180, -200, -201; PA-28RT-201); Symbolic Displays, Inc., 1762 McGaw Ave., Irvine, CA 92714.

SA148RM: Install SDI Model CFS-1000A, -1001A, FT-100, or FT-101 fuel-flow indicating system (PA-28-201T; PA-28R-201T; PA-28RT-201T); Symbolic Displays, Inc., 1762 McGaw Ave., Irvine, CA 92714.

SA3771WE: Installation of Silver Instruments Fueltron IG or IP or Fuelgard digital fuel-flow and totalizer system (PA-28R-201T); Silver Instruments, Inc., 8202 Capwell Dr., Oakland Airport Business Park, Oakland, CA 94621.

SA4459NM: Installation of Silver Instruments Fueltron IG-C and IG-SC and Fuelgard digital fuel-flow indicating system (PA-28-236); Silver Instruments, Inc., 8202 Capwell Dr., Oakland Airport Business Park, Oakland, CA 94621.

PA-32 Series

SA311NM: Installation of flap extension, spoiler L/E cuff and optional cambered wingtips (PA-32); Robertson Aircraft, 839 W. Perimeter Rd., Renton, WA 98055.

SA3NW: Installation of Van's Aircraft Model VA-1 fiberglass wheel fairings (PA-32-260, 300); Van's Aircraft, Route 2 Box 187, Forest Grove, OR 97116.

SA530CE: Install Fluidyne Model 4000 and 2500A skis (PA-32-260, 300); Fluidyne Engineering Corp, 5900 Olson Memorial Hwy, Minneapolis, MN 55422.

SA1568SW: Wing leading edge cuff, flow fences, optional wingtips and dorsal fin (PA-32-260, -300); Barbara or Bob Williams, Box 431, 213 North Clark, Udall, KS 67146.

SA2217WE: Installation of recontoured wing leading edge, raked wingtips, fuselage flap, stall fences, droop ailerons and dorsal fin (optional) (PA-32-260, -300); Charyl C. Robertson, 1540 Sunset Hwy, Bellevue, WA 98007.

SA1486SO: Installation of speed enhancement kit (PA-32-260, -300, -301; PA-32R-300, -300T, -301, -301T; PA-32RT-300, -301T); Sea Wings, Inc., 1-1 Wintberg Skyline Dr., St. Thomas, VI.

SA820GL: Installation of aileron and flap gap seals (PA-32-260, -300; PA-32R-300); General Aviation Corp., Rock County Airport, Janesville, WI 53545.

SA609GL: Installation of aileron, flap, and stabilator gap seals (PA-32-260, -300; PA-32R-300; PA-32RT-300, -300T); Knots 2 U, Inc., 1941 Highland Ave., Wilmette, IL 60091.

SA4288WE: Installation of inflatable door seal (PA-32 series); Bob Fields Aerocessories, P.O. Box 390, Santa Paula, CA 93060.

SA932EA: Installation of Pee Kay Model B3500 seaplane floats and Pee Kay Model 3500A amphibious floats (PA-32S-300); Pee Kay DeVore, Inc., 125 Mineola Ave., Roselyn Heights, NY 11577.

SA2253WE: Installation of Edo Model FD-3500-21 amphibious floats (PA-32S-300); T.M. Close Corp., Gardner Municipal Airport, Box 464, Gardner, MA 01440.

SA409GL: Installation of a Lycoming IO-540-K1A5 engine, Hartzell propeller, and associated components (PA-32-260); Melvin C. Morkert, 915 Montclair Dr., Racine, WI 53402.

SA1557WE: Installation of turbocharged Lycoming O-540-E4B5 engine (PA-32-260); Roto-Master, Inc., 7101 Fair Ave., North Hollywood, CA 91605.

SA3736WE: Installation of an engine oil cooler hose (PA-32-260, -300; PA-32R-300); Aircraft Metal Products, Inc., 4206 Glencoe Ave., Venice, CA 90291.

SA3839WE: Installation of an air/oil separator (PA-32-260, -300;

PA-32R-300); Walker Engineering Co., P.O. Box 8151, Van Nuys, CA 91409.

SA893EA: Installation of eight-quart alcohol propeller anti-icing kit WAP-800A (PA-32-300); E.W. Wiggins Airways, Inc., Norwood Municipal Airport, Norwood, MA 02062.

SA3513WE: Installation of Rajay turbocharged Lycoming IO-540-K1G5D engine (PA-32R-300); Rajay Industries, Inc., 2600 East Wardlow Rd., P.O. Box 207, Long Beach, CA 90801.

SA4345WE: Installation of cooling louvers in the top cowling (PA-32R-300T); Marina Spear, 5555 Corso di Napoli, Long Beach, CA 90803.

SA371EA: Zeiss aerial camera and intervalometer (PA-32-260); James W. Sewall Co., 147 Center St., Old Town, ME 00468.

SA1310SW: Stretcher in lieu of rear seats (PA-32-260); Grand Prairie Flying Service, Almyra Airport, Box 220, Almyra, TX 72203.

SA1362CE: Chrome-plated brake disc installation (PA-32-260, -300); Engineering Plating and Processing, Inc., 641 Southwest Blvd., Kansas City, KS 66103.

SA2933WE: Installation of oxygen system (PA-32-260, -300); Sky Ox Limited, P.O. Box 6, Saint Joseph, MI 49085.

SA4327WE: Installation of rudder, brake, and flap systems hand controls (PA-32-260, -300, -301, -301T); Terry Doty, 19146 San Jose Ave., La Puente, CA 91748.

SA21NE: Installation of Whelen A600-PR and A600-PG anti-collision strobe with forward and tail position-light assemblies as replacements for wingtip lights (PA-32R-300); Whelen Engineering Co., Inc., Winter Ave., Deep River, CT 06417.

SA1159EA: Installation of DeVore "Tel-Tail" lights on lower surface of horizontal stabilator (PA-32RT-300, -300T); DeVore Aviation Corp., 6104-B Kircher Blvd, NE, Albuquerque, NM 87109.

SA836GL: Installation of a fuel-flow meter and totalizer system (PA-32 series); Shadin Company, Inc., 6950 Wayzata Blvd, Suite 221, Minneapolis, MN 55426.

SA3670WE: Installation of Silver Instruments Fueltron IG or IP or Fuelgard digital fuel-flow indicating system (PA-32-300, -301; PA-32R-300, -301); Silver Instruments, Inc., 1896 National Ave., Hayward, CA 94545.

SA3774WE: Installation of SDI Model CFS-1000, 1001, FT-100, or FT-101 fuel-flow indicating system (PA-32-300, -301; PA-32R-300, -301; PA-32RT-300, -301); Symbolic Displays, Inc., 1762 McGaw Ave, Irvine, CA 92705.

SA975SO: Installation of two 15-gallon auxiliary fuel tanks, one in each wing (PA-32R-300); Cypress Aviation, Inc., 3480 Drane Field Rd., Lakeland, FL 33803.

PA-38 Series

SA2059NM: Installation of a modified Lycoming O-235-L2C designated O-235-L2C(M) in accordance with STC792NW 8PA-38-112; Aeromod Corp., Bldg. C3, Paine Field, Everett, WA.

Contact the STC holders for further information about their products.

POPULAR MODIFICATIONS

For the pilot going places, speed is always important. Speed is also an indicator of efficiency, which affects direct operating costs. To improve aircraft efficiency, a modification or smoothing of the airflow around the airplane can be made.

These modifications include gap seals, wheel pants, and other items designed to smooth the airflow (Figs. 9-1 through 9-6).

For further information about these and similar modifications, contact:

Knots 2 U, Inc.
1941 Highland Ave.
Wilmette, IL 60091
Phone: (312) 256-4807

Univair
2500 Himalaya Rd.
Aurora, CO 80111
Phone: (303) 364-7661

Wag-Aero, Inc.
Box 181
Lyons, WI 53148
Phone: (414) 763-9586

Fig. 9-1. Wheel pants are perhaps the most popular modifications made for the PA-28 and PA-32 fixed-gear airplanes.

Fig. 9-2. Notice the open rudder gap on the left and the ''sealed'' gap on the right. Fewer gaps mean smoother airflow and greater speed. (courtesy Knots-2-U)

Fig. 9-3. This is a complete gap seal kit for flaps and ailerons. (courtesy Knots-2-U)

Fig. 9-4. After the kit has been installed, the air can flow smoothly over the wing. (courtesy Knots-2-U)

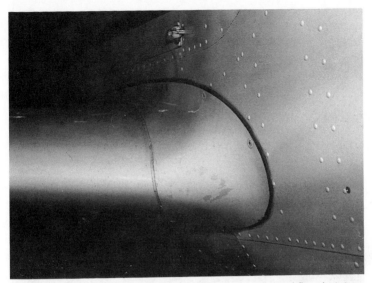

Fig. 9-5. Even the wing roots must have fairings to smooth the airflow. (courtesy Knots-2-U)

Fig. 9-6. A popular modification to the Comanche is the installation of this much larger windshield, which gives improved visibility. (courtesy Knots-2-U)

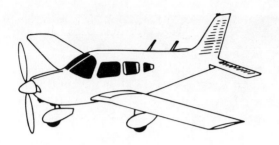

Chapter 10
Buying a Used Airplane

The search for a good, used, all-metal Piper does not usually have to be wide and exhaustive. Simply by sheer numbers, there are many from which to choose. This is good for the buyer (Fig. 10-1).

BEFORE MAKING A COMMITMENT

When thinking about the use your airplane will get, you must be very objective and completely honest with yourself. Remember, you are going to have this airplane for a long time. You must be *very* satisfied with it.

How many people fly with you? Is most of your flying done by yourself, or perhaps with your wife or a flying buddy? If this is so, then perhaps a two-place airplane would suit you. Otherwise, look for a four-place craft.

How far do you go on the average flight? Are most of your flights under 100 miles, or do you frequently fly hundreds of miles on business? Remember, be objective.

Closely related to distance is the question of speed. How fast do you *really* need to get there? If you are going to fly for business, there may be need for a fast airplane. However, remember that speed usually equates to larger engines, more complex airplanes, and the resulting higher operation and maintenance expenses.

Avionics add much to the value of a used airplane. They also add to the maintenance that keeps those ''black boxes'' working.

If you need them, then by all means have them. If you are not instrument-rated, or don't fly in bad weather, then you don't need them. Save your money. Only buy what you need.

The airports that you will operate from will influence your choice of airplanes. If all of your flying is from paved runways, then you have many possibilities. However, if you are flying from rough grass strips or unimproved areas, then you must select an airplane that will stand up to the abuse.

Do you fly for sport (evenings and weekends), transportation (business and family outings), or hauling (carrying light cargo)?

Maintenance is also a major consideration when purchasing an airplane. There are annual inspections (required by law), minor repairs, and major repairs—many places for your dollars to go. There is also the Airworthiness Directive. An AD, as explained in Chapter 8, is a maintenance step *required* by the FAA. Some of them call for extra inspections of airframe/engine parts; others call for changes to the system. They are all done in the name of safety, and they all call for additional dollars to be spent.

Maintenance should weigh heavily in your decision-making. The more complex an airplane you purchase, the more maintenance dollars you will spend. Consider how maintenance-free the following choices could be:

☐ Landing gear: Fixed vs. retractable

☐ Propeller: Fixed vs. variable pitch

☐ Avionics: VFR vs. IFR

With just these three comparative differences, you should be able to understand that simple means *less expensive*. I'll not use the word *cheap*, as there is nothing cheap about airplane maintenance. Everything is expensive; some things are just less expensive than others. In an airplane, if you don't have it, it can't break; therefore, it will cost nothing to maintain.

But remember, the final objective is to select an airplane that will fill your needs, fulfill your desires, be affordable to own and operate, and not overtax your piloting skills.

Don't buy more airplane than you need or can handle. If you can't handle it, you won't enjoy it, and therefore you won't fly it.

THE SEARCH

As there are many models and price ranges to select from, the purchaser is encouraged to set a range of his expectations. This

Fig. 10-1. Somewhere, an airplane waits for you. The problem is to find it!

should be based upon features desired, options available, or, most likely, the cash available for such a purchase.

The search should start locally, as this will make it easy for you to see what is available on the market. Usually you should start looking on your home field. If you know the FBO, and feel comfortable with him, perhaps you should enlist his aid in your search. Discuss your desires with him. Often an FBO will know of airplanes for sale, or nearly for sale, that have yet to be advertised. After all, he is an insider to the business.

If there is nothing of interest at your field, then broaden the search. Check the bulletin boards at other nearby airport(s). Ask around while you're there, then walk around and look for airplanes with "For Sale" signs in the windows. You could even put an "Airplane Wanted" ad on the bulletin boards.

The local newspaper will sometimes have airplanes listed in their classified ads. However, in this day of specialization, two organizations have become leaders in airplane advertising:

Trade-A-Plane
Crossville, TN 38555

Air Show
Suite 205
45 West Broadway
Eugene, OR 97401
Phone: (800)247-9005
      (503)344-7813

Trade-A-Plane is the oldest of all printed airplane advertising. For $10.00 you will receive three issues each month for six months. Each issue will provide enough yellow pages of newsprint to keep you reading more than a single evening—and wallpaper the inside of any one-plane hangar. *Trade-A-Plane* has long been considered to be the single source of all aviation products. Anyone owning, or considering ownership, should read it.

Air Show, an aircraft marketing firm, has recently introduced a complete sales system designed to aid the long-distance buyer. Air Show lists aircraft for sale in the *Air Show Journal*, their twice-monthly publication, and in other national-coverage "airplane" papers, magazines, and *Trade-A-Plane*. The *Air Show Journal* is directed toward general aviation.

The *Journal* includes news and informational articles as well as aircraft advertising. Air Show provides an "800" toll-free telephone number, which makes the search less expensive. However, more than just advertise, Air Show offers a unique means of complete visual inspection of an airplane with no need for long-distance travel.

Air Show offers a videotaped 15-minute documentary about each airplane they list for sale. For a small fee, the prospective purchaser can have a copy of this sales tape sent via UPS, and make a complete "used airplane inspection" in the privacy and convenience of his own home.

These video recordings follow the same general walk-around inspection recommended later in this chapter. The tapes are, for the most part, filmed outside, and not under "studio" conditions. The net result is a very informative video representation of the aircraft being "inspected."

Other listings of used airplanes can be found in the various flight-oriented magazines (i.e., *AOPA Pilot*, *FLYING*, *Plane & Pilot*, *Private Pilot*, etc.).

Most ads of airplanes for sale make use of various more-or-less standard abbreviations. These abbreviations describe the individual airplane and tell how it is equipped. Also in the ads will be a telephone number, but seldom a location of where the airplane is located (a clue is the area code).

Here's a sample ad:

67 Cherokee 180,3189TT,675 SMOH,Apr ANN,FGP,
Dual NAV/COM,GS,MB,ELT,NDH. $13,750 firm.
703-555-1212

Translated, this ad reads: For sale, a 1967 Piper PA-28 Cherokee Model 180. The airplane has a total of 3189 hours on the
airframe, and an engine with 675 hours since a major overhaul. The
next annual inspection is due in April. It is equipped with a full gyro
instrument panel, has two navigation and communication radios, a
glideslope receiver and indicator, a marker beacon receiver, an
Emergency Locator Transmitter, and best of all, the airplane has
no damage history. The price is $13,750, and the seller claims he
will not bargain. (Most do, however.) Lastly is the telephone number.

As you can see, there sure was a lot of information inside those
three little lines.

Advertising Abbreviations

A/C	air conditioning
AD	Airworthiness Directive
ADF	Automatic Direction Finder
A&E	airframe and engine
AF	airframe
AF&E	airframe and engine
AI	Aircraft Inspector
ALT	altimeter
ANN	annual inspection
ANNUAL	annual inspection
AP	autopilot
A/P	autopilot
ASI	airspeed indicator
BAT	battery
CAT	carburetor air temperature
CHT	cylinder-head temperature
COMM	communications radio
CS	constant-speed propeller
C/S	constant-speed propeller
C/W	complied with
DBL	double
DG	directional gyro
DME	Distance Measuring Equipment
FAC	factory
FBO	fixed base operator
FGP	full gyro panel
FWF	firewall forward

GAL	gallons
GPH	gallons per hour
GS	glide slope
HD	heavy-duty
HP	horsepower
HSI	horizontal situation indicator
HVY	heavy
IFR	Instrument Flight Rules
ILS	Instrument Landing System
INSP	inspection
INST	instrument
KTS	knots
L	left
LDG	landing
LE	left engine
LED	light-emitting diode
LOC	localizer
LTS	lights
MB	marker beacon
MBR	marker beacon receiver
MGTW	maximum gross takeoff weight
MOD	modification
MP	manifold pressure
MPH	miles per hour
MTOW	maximum take-off weight
NAV	navigation
NAV/COM	navigation/communication radio
NDH	no damage history
OAT	outside air temperature
OX	oxygen
O2	oxygen
PAX	passengers
PROP	propeller
PSGR	passenger

PSI	pounds per square inch
R	right
RC	rate of climb
RE	right engine
REMAN	remanufactured
REPALT	reporting altimeter
RMFD	remanufactured
RMFG	remanufactured
RNAV	Area Navigation
ROC	rate of climb
SAFOH	since airframe overhaul
SCMOH	since (chrome/complete) major overhaul
SFACNEW	since factory new
SFN	since factory new
SFNE	since factory new engine
SFREM	since factory remanufacture
SFREMAN	since factory remanufacture
SFRMFG	since factory remanufacture
SMOH	since major overhaul
SNEW	since new
SPOH	since propeller overhaul
STC	Supplemental Type Certificate
STOH	since top overhaul
STOL	short takeoff and landing
TAS	true airspeed
T&B	turn and bank
TBO	time between overhaul
TC	turbocharged
TLX	telex
TNSP	transponder
TNSPNDR	transponder
TSN	time since new
TSO	Technical Service Order
TT	total time
TTAF	total time airframe
TTA&E	total time airframe and engine

TTE	total time engine
TTSN	total time since new
TXP	transponder

VAC	vacuum
VFR	Visual Flight Rules
VHF	very high frequency
VOR	Very High Frequency Omni Range

XC	cross-country
XLNT	excellent
XMTR	transmitter
XPDR	transponder
XPNDR	transponder

| YD | yaw damper |

3LMB	three-light marker beacon
3BL	three-blade propeller
3BLP	three-blade propeller

Area Codes

Some advertisements contain a telephone number. By locating the telephone area code region, you can eliminate airplanes that are too far away to look at, or from undesirable coastal areas.

201	NJ north	214	TX Dallas
202	Washington, DC	215	PA east
203	CT	216	OH northeast
205	AL	217	IL central
206	WA west	218	MN north
207	ME	219	IN north
208	ID	301	MD
209	CA Fresno	302	DE
212	NY City	303	CO
213	CA Los Angeles	304	WV

305	FL southeast	609	NJ south
307	WY	612	MN central
308	NE west	614	OH southeast
309	IL Peoria area	615	TN east
312	IL northeast	616	MI west
313	MI east	617	MA est
314	MO east	618	IL south
315	NY north central	619	CA southeast
316	KS south	701	ND
317	IN central	702	NV
318	LA west	703	VA north and west
319	IA east	704	NC west
401	RI	707	CA Santa Rosa area
402	NE east	712	IA west
404	GA north	713	TX Houston
405	OK west	714	CA southwest
406	MT	715	WI north
408	CA San Jose area	716	NY west
409	TX southeast	717	PA central
412	PA southwest	718	NY southeast and NYC
413	MA west	801	UT
414	WI east	802	VT
415	CA San Francisco	803	SC
417	MO southwest	804	VA southeast
419	OH northwest	805	CA west central
501	AR	806	TX northwest
502	KY west	808	HA
503	OR	812	IN south
504	LA east	813	FL southwest
505	NM	814	PA northwest and central
507	MN south	815	IL north central
509	WA east	816	MO northwest
512	TX south central	817	TX north central
513	OH southwest	818	CA southwest
515	IA central	901	TN west
516	NY Long Island	904	FL north
517	MI central	906	MI northwest
518	NY northeast	907	AK
601	MS	912	GA south
602	AZ	913	KS north
603	NH	914	NY southeast
605	SD	915	TX southwest
606	KY east	916	CA northwest
607	NY south central	918	OK northeast
608	WI southwest	919	NC east

THE INSPECTION

The object of the pre-purchase inspection of a used airplane is to preclude the purchase of a "dog." No one wants to buy someone else's troubles. The pre-purchase inspection must be completed in an orderly manner. Take your time during this inspection; a few extra minutes spent inspecting could well save you thousands of dollars later.

The very first item of inspection is a question directed to the current owner: "Why are you selling it?" Of course, if the seller has something to hide, you may not get an honest answer. Fortunately, most people will answer honestly. Often the owner is moving up to a larger plane, and if so, he will start to tell you about his new prospective purchase. Let him talk; you can learn a lot about him by listening. You can gain insight into his flying habits and how he treated the plane you are considering purchasing. Perhaps he has other commitments (i.e., spouse says sell, or perhaps he can no longer afford the plane). Financial or family pressures can work to your advantage; however, there is a warning here also. If the seller is having financial difficulties, consider the quality of maintenance that was performed on the airplane.

Ask the seller if he knows of any problems or defects with the airplane. Most sellers will give you their honest answer, but there could be things he doesn't know about.

Buyer Beware—it's *your* money and *your* safety!

Definitions to Know

Airworthy: The airplane must conform to the original type certificate, or those Supplemental Type Certificates (STCs) issued for this particular airplane (by serial number). In addition, the airplane must be in safe operating condition (relative to wear and deterioration).

Annual Inspection: All small airplanes must be inspected annually by an FAA-certified Airframe & Powerplant mechanic who holds an IA (Inspection Authorization), by an FAA-certified repair station, or by the airplane's manufacturer. This is a complete inspection of the airframe, powerplant, and all subassemblies.

100-Hour Inspection: This is of the same scope as the annual, and is required on all commercially operated small airplanes (i.e., rental, training, etc.), and must be accomplished after every 100 hours of operation. This inspection may be performed by an FAA-certified Airframe & Powerplant mechanic without an IA rating. An annual inspection will fulfill the 100-hour requirement, but the reverse is not true.

Pre-flight Inspection: A thorough inspection, by the pilot, of an aircraft prior to flight. The purpose is to spot obvious discrepancies by inspection of the exterior, interior, and engine of the airplane.

Preventive Maintenance: FAR Part 43 lists a number of maintenance operations that are considered preventive in nature and may be performed by a certificated pilot on an airplane he owns, provided the airplane is not flown in commercial service. (These operations are described in Chapter 12 of this book.)

Repairs and Alterations: There are two classes of repairs and/or alterations: *Major* and *Minor*.

Major repairs/alterations must be approved for a return to service by an FAA-certified Airframe & Powerplant mechanic holding an IA rating, by a repair station, or by the FAA.

Minor repairs/alterations may be returned to service by an FAA-certified Airframe & Powerplant mechanic.

Airworthiness Directives: ADs are defined in FAR Part 39, and must be complied with. As described in Chapter 8, they are *required* maintenance/repair procedures.

Files of ADs and their requirements are kept by mechanics and FAA offices. A compliance check of ADs is a part of the annual inspection.

Service Difficulty Reports: SDRs are prepared by the FAA from Malfunction or Defect Reports (MDRs) that are initiated by owners, pilots, and mechanics. SDRs are not the word of law that ADs are; however, they should be adhered to for your own safety.

TBO: The "time between overhauls" recommended by the manufacturer as the maximum engine life. It has no legal bearing on airplanes not used in commercial service; it's only an indicator. Many well-cared-for engines last hundreds of hours beyond TBO—but not all.

Remanufacture: The complete disassembly, as-needed repair, alteration/updates, and inspection of an airplane engine. This includes bringing all engine specifications back to factory-new limits. A factory-remanufactured engine comes with new logs and zero time. FARs state that only the engine manufacturer or a factory-approved agency may "zero" an engine. The factory-remanufactured engine is the next thing to a new engine.

There is considerable confusion about the terms *remanufactured* and *rebuilt*. Rebuilt and remanufactured have become synonymous over the past few years, no doubt as an industry move to confuse the aircraft owner. In FAR 43.2(b) and 91.175 are prohibitions against using the term "rebuilt" in describing engines. 91.175 refers to *factory-rebuilt* engines, and defines them the same as *remanufactured* (above). It has become common practice to call a factory-rebuilt engine *factory remanufactured*. Only the factory can rebuild (remanufacture) an aircraft engine, and give a "zero time" log.

overhaul: The disassembly, inspection, cleaning, repair, and reassembly of an airplane engine. The work may be done to new limits or to *service limits*.

top overhaul: The rebuilding of the head assemblies, but not of the entire engine. In other words, the case of the engine is not split, only the cylinders are pulled. The top overhaul is utilized to bring an oil-burning and/or low-compression engine within specifications. It is a method of stretching the life of an otherwise sound engine. A top overhaul can include such work as valve replacement/grinding, cylinder replacement/repair, piston and ring replacement, etc. It is not necessarily an indicator of a poor engine. The need for a "top" may have been brought on by such things as lack of care, lack of use of the engine, or plain abuse (i.e., hard climbs and fast letdowns). An interesting note: The term *top overhaul* does not indicate the extent of the rebuild job (i.e., number of cylinders rebuilt or the completeness of the job).

new limits: The dimensions/specifications used when constructing a new engine. Parts meeting new limits will normally reach TBO with no further attention, except for routine maintenance.

service limits: The dimensions/specifications below which use is forbidden. Many used engine parts will fit into this category; however, they are unlikely to last the full TBO, as they are already partially worn.

nitriding: A method of hardening cylinder barrels and crankshafts. The purpose is to create a hard surface that resists wear, thereby extending the useful life of the part.

chrome plating: Used to bring the internal dimensions of the cylinders back to specifications. It produces a hard, machinable, and long-lasting surface. There is one major drawback of chrome plating: longer break-in times. However, an advantage of chrome plating is its resistance to destructive oxidation (rust) within combustion chambers.

THE VISUAL INSPECTION

The visual inspection of a used airplane is basically a very thorough preflight, with a few extras included. It's divided into four simple, yet logical, steps.

The Cabin

Open the door and look inside. Notice the general condition of the interior. Does it appear clean, or has it just been scrubbed after a long period of inattention? Look in the corners, just as you would if you were buying a used car. Does the air smell musty and damp? Is the headliner in one tight piece, and the upholstery unfrayed? In what condition are the door panels? The care given the interior of an airplane can be a good indication of what care was given to the remainder of the airplane. If you have purchased used cars in the past, and have been successful, you are qualified for this phase of the inspection.

Look at the instrument panel. Does it have what you want and need? Are the instruments in good condition, or are there knobs missing and glass faces broken? Is the equipment all original, or have there been updates made? If updates have been made, are they neat in appearance—and workable? Often, updating—particularly in avionics—is done haphazardly, with results that are neither pleasing to the eye, nor workable to the pilot.

Look out the windows. Are they clear, unyellowed, and uncrazed? Side windows are not expensive to replace, and you can do it yourself. Windshields are another story—and another price.

Check the operation of the doors. They should close and lock with little effort; no outside light should be seen around edges of the doors.

Check the seats for freedom of movement and adjustability. Check the seat tracks and the adjustment locks for damage.

Airframe

Do a walk-around and look for the following:

Is the paint in good condition, or is some of it lying on the ground under the airplane? Paint jobs are expensive, yet necessary for the protection of the metal surfaces from corrosive elements. Paint jobs also please the eye of the beholder. A good paint job will cost in excess of $2,500.

Dents, wrinkles, or tears of the metal skin can indicate prior damage, or just careless handling. Each discrepancy must be examined very carefully by an experienced mechanic. Total consideration of all the dings and dents will indicate if the airplane has had an easy or a rough life.

Corrosion or rust on skin surfaces or control systems should be cause for alarm. Corrosion is to aluminum what rust is to iron. It's destructive. Any corrosion or rust should be brought to the attention of a mechanic for his judgment. Corrosion that appears as only minor skin damage can continue, unseen, into the interior structure. Corrosion such as this creates dangerous structural problems that can be very costly to repair. Having a ''little'' corrosion in the wing structure is similar to having a ''little'' lung cancer. The word ''little'' is irrelevant.

Be alert for extensive corrosion in aircraft from coastal areas. Salt and aluminum don't get along well.

Check for fuel leaks around the wings—in particular, where the wings attach to the fuselage. If leakage evidence is seen, have a mechanic check its source.

The landing gear should be checked for evidence of having been sprung. Check the tires for signs of unusual wear that might indicate other structural damage. Also, look at the oleo struts for signs of fluid leakage and proper extension.

Move all the control surfaces, and check each for damage. They should be free and smooth in movement. When the controls are centered, the surfaces should also be centered. If they are not centered, a problem in the rigging of the airplane exists.

Engine

Open or remove the cowling to inspect the engine. If you cannot see the engine, you cannot inspect it! Search for signs of oil leakage. Do this by looking at the engine, the inside of the cowl, and the firewall. If the leaks are bad enough, there will be oil dripping to the ground or onto the nosewheel. Naturally, the seller has probably cleaned all the old oil drips away; however, oil leaves stains. Look for these stains.

Check all the fuel and vacuum hoses/lines for signs of deterioration or chafing. Also, check the connections for tightness and/or signs of leakage.

Check control linkages and cables for obvious damage and ease of movement. Be sure none of the cables is frayed.

Check the battery box and battery for corrosion.

Check the propeller for damage, such as nicks, cracks, or gouges. Even very small defects can cause stress areas on the prop (see Chapter 14). Any visible damage to a propeller must be checked by a mechanic. Also check it for movement that would indicate propeller looseness at the hub.

Check the exhaust pipes for rigidity, then reach inside them by rubbing your finger along the inside wall. If your finger comes back perfectly clean, you can be assured that someone has cleaned the inside of the pipe—possibly to remove the oily deposits that form there when an engine is burning a lot of oil. If your fingers come out of the pipe covered with a black oily goo, have your mechanic determine the cause. It might only be a carburetor in need of adjustment, but it could also be caused by a large amount of oil blow-by, the latter indicating an engine in need of large expenditures for overhaul. A light gray dusty coating indicates proper operation.

Check for exhaust stains on the belly of the plane to the rear of the exhaust pipe. This area has probably been washed, but look anyway. If you find black oily goo, then, as above, see your mechanic.

Logbooks

If you are satisfied with what you've seen up to this point, then go back to the cabin, have a seat, and check that all of the required paperwork is with the airplane. This includes:

☐ Airworthiness Certificate

☐ Aircraft Registration Certificate

☐ FCC Station License

☐ Flight Manual or operating limitations

☐ Logbooks (airframe, engine, and propeller)

☐ Current equipment list

☐ Weight and balance chart

These items are required by the FARs to be in the plane (except for the logs, which must be available).

Pull out the logbooks and start reading them. Sitting there will also allow you to look, once again, around the cockpit.

Be sure you're looking at the proper logs for this particular aircraft, and that they are the original logs. Sometimes logbooks get "lost" and are replaced with new ones. This can happen because of carelessness or theft. This is why many owners do not keep their logs in the plane, and may only provide copies for a sales inspection. Replacement logs may be lacking very important information, or could be outright frauds. Fraud is not unheard of in the used airplane business. Be on your guard if the original logs are not available.

Check the airplane's indentification plate for serial and model numbers. The ID plate is on the left side of the fuselage, at the rear of the plane. The numbers must match with the logbooks. The year of manufacture can be determined by reference to the serial number lists in earlier chapters of this book.

Start with the airframe log by looking in the back for the AD compliance section (see Chapter 8 for a list of the ADs). Check that the list is up-to-date, and that any required periodic inspections have been made. Now go back to the most recent entry. It probably is an annual or 100-hour inspection. The annual inspection will be a statement that reads:

October 29, 1987   Total Time: 2,978 hrs.

I certify that this aircraft has been inspected in accordance with an annual inspection and was determined to be in airworthy condition.

signed here

IA # 0000000

From this point back to the first entry in the logbook, you'll be looking for similar entries, always keeping track of the total time, for continuity purposes and to indicate the regularity of usage (i.e., number of hours flown between inspections). You will also be looking for indications of major repairs and modifications. This will be signaled by the phrase, "Form 337 filed." A copy of this form should be with the logs, and will tell what work was done. The work may also be described in the logbook (Fig. 10-2).

Form 337, Major Repair and Alteration, is filed with the FAA, and copies are a part of the official record of each airplane. They are retrievable from the FAA, for a fee.

The engine log will be quite similar in nature to the airframe log, and will contain information from the annual and/or 100-hour inspections. Total engine time will be given, and possibly an indication of time since any overhaul work, although you may have to do some math here. It's quite possible that this log—and engine—will not be the original for the aircraft. As long as the facts are well-documented in both logs, there is no cause for alarm. This would be the case if the original engine was replaced with a factory-rebuilt one, or even a used engine from another plane.

Pay particular attention to the numbers that indicate the results of a differential compression check. These numbers are the best single indicator of the overall health of an engine.

Each number is given as a fraction, with the denominator always being 80. The 80 indicates the air pressure that was utilized for the check (80 pounds per square inch is the industry standard). The numerator is the air pressure that the combustion chamber was able to maintain while being tested—80 would be perfect, but it isn't attainable; it will always be less. The reason for the lower number is the air pressure loss that results from loose, worn, broken rings; scored or cracked cylinder walls; or burned, stuck, or poorly seated valves. There are methods mechanics use to determine which of these is the cause and to repair the damage.

Normal readings would be no less than 70/80, and should be uniform (within 2-3 psi) for all cylinders. A discrepancy between cylinders could indicate the need for a top overhaul of one or more cylinders. The FAA says that a loss in excess of 25 percent is cause for further investigation (that would be a reading of less than 60/80). I personally feel such a low reading indicates a very tired engine, one in need of considerable work and expenditures.

Read the information from the last oil change. It may contain a statement about debris found on the oil screen or in the oil filter. However, oil changes are often performed by owners, and may or may not be recorded in the log, even though the FARs require all maintenance to be logged. If the oil changes are recorded, how regular were they? I prefer every 25 hours, but 50 is the norm. Is there a record of oil analysis available? If so, ask for it.

If the engine has been top overhauled or majored, there will be a description of the work performed, a date, and the total time on the engine when the work was accomplished.

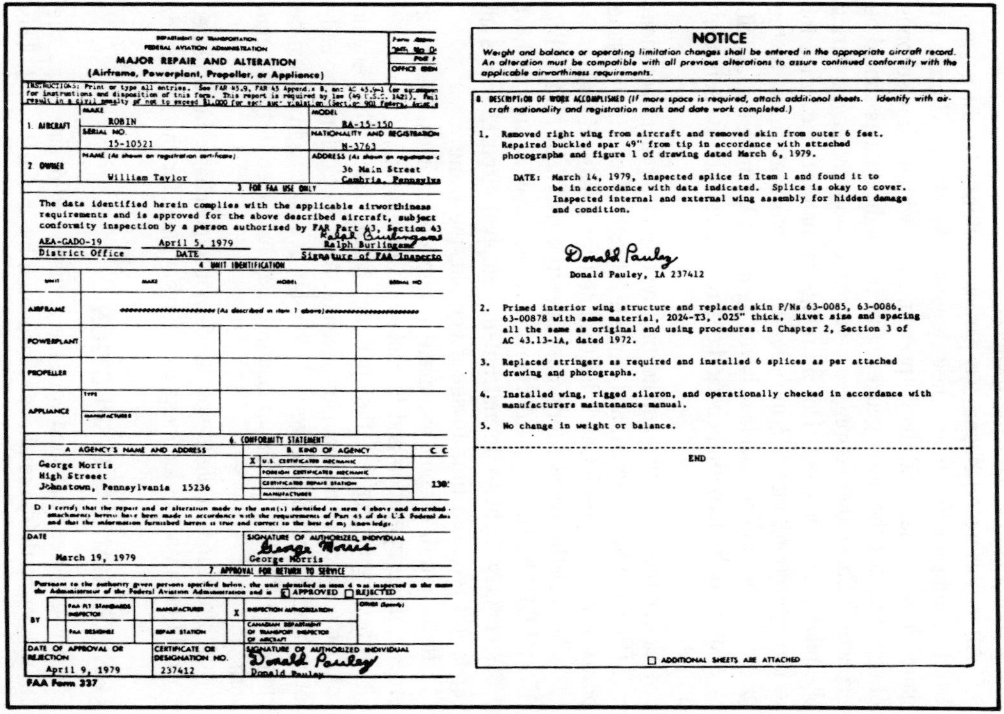

Fig. 10-2. FAA Form 337 can contain much important information about work performed on an aircraft. (courtesy FAA).

Check to see if the ADs have been complied with, and the appropriate entries made in the log (Fig. 10-3).

THE TEST FLIGHT

The test flight is to determine if the airplane "feels" right to you. The flight should last at least 30 minutes, but two hours would not be too much.

For insurance purposes, I recommend that either the owner or a competent flight instructor accompany you on the test flight. This will also eliminate problems of currency, ratings, etc., and it will foster better relations with the owner.

After starting the engine, pay particular attention to the gauges. Do they jump to life, or are they sluggish? Watch the oil pressure gauge in particular. Did the oil pressure rise within a few seconds of start? Check the other gauges. Are they indicating as should be expected? Check them during your ground run-up, then again during the takeoff and climbout. Do the numbers match those called for in the operations manual? In order to pay more attention to the gauges, it might be advisable to have the other pilot make the takeoff.

After you're airborne, check the gyro instruments. Be sure they are stable.

Check the ventilation and heating system for proper operation.

Do a few turns, stalls, and some level flight. Does the airplane perform as expected? Can it be trimmed for hands-off flight?

Check all the avionics for proper operation (NAV/COMM, MBR, ADF, LORAN, ILS, etc.) A complete check may require a short cross-country flight to an instrument-equipped airport. That's all right; it'll give you time to see if you like the plane.

Return to the airport and make a couple of landings. Check for proper brake operation and for nosewheel shimmy.

After returning to the parking ramp, open the engine compartment and look again for oil leaks. Also check along the belly for indications of oil leakage and blow-by. A short flight should be enough to "dirty" things up again, if they had been dirty to begin with.

If, after the test flight, you decide not to purchase the airplane, it would be an act of kindness to offer payment for the fuel used.

THE MECHANIC'S INSPECTION

If you are still satisfied with the airplane and desire to pursue the matter further, then have it inspected by an A&P or AI. This

APPENDIX 1. AIRWORTHINESS DIRECTIVE COMPLIANCE RECORD

*Aircraft, Engine, Propeller, Rotor, or Appliance: Make _____ Model _____ Ser.No. _____ N _____

AD Number and Amendment Number	Date Received	Subject	Compliance Due Date Hours/ Other	Method of Compliance	Date of Compliance	Airframe Total Time In Service at Compliance	Component Total Time In Service at Compliance	One-Time	Recur-ring	Next Comp. Due Date Hours/ Other	Authorized Signature, Certificate Type and Number	Remarks

*Suggest providing a page for each category.

Fig. 10-3. The suggested format for an AD compliance log. (courtesy FAA)

177

inspection will cost you a few dollars; however, it could save you thousands. The average for a pre-purchase inspection is three to four labor hours at shop rates.

The mechanic will accomplish a search of ADs, a complete check of the logs, and an overall check of the plane. A compression check and a borescope examination must be made to determine the internal condition of the engine. A borescope examination means looking into a cylinder and viewing the top of the piston, the valves, and the cylinder walls. This is done by use of a special device called—naturally—a borescope.

Always use your own mechanic for the pre-purchase inspection. By this, I mean someone *you* are paying to watch out for *your* interests, not someone who may have an interest in the sale of the plane (i.e., employee of the seller).

Have the plane checked even if an annual was just done, unless you know and trust the AI who did the inspection. You might be able to make a deal with the owner over the cost of the mechanic's inspection, particularly if an annual is due.

It's not uncommon to see airplanes listed for sale with the phrase "annual at date of sale." I am always leery of this, because I don't know who will do the annual, or how complete it will be. All annuals are not created equal! An "annual at date of sale" is coming with the airplane, done by the seller as part of the sale. Who is looking out for *your* interests?

MOST IMPORTANT

In each of my books, when I talk about purchasing used airplanes I always give the following advice:

If an airplane seller refuses to let you do anything that has been mentioned in this chapter, then thank him for his time, walk away, and look elsewhere. Do not let a seller control the situation. Your money, your safety, and possibly your very life are at stake. Airplanes are not hot sellers, and there is rarely a line forming to make a purchase. *You* are the buyer; *you* have the final word.

In retrospect, after a few sales pitches I've recently heard, I would make the thanks optional!

WHAT VALUE? WHAT PRICE?

You have decided this airplane is it—you just can't do without it. All inspections have been made, and you are satisfied the airplane will suit your needs. Is the price agreeable? Does the asking price equal the value?

The value of an aircraft is controlled by many factors:

☐ The age and general condition of the airplane

☐ How well equipped the plane is, and the age/condition of that equipment

☐ The history of the aircraft, and its past usage and damage record

☐ AD requirements and compliance

☐ The remaining time left on limited-life components (engine, propellers, etc)

Engine Values

The time on an engine, since new or overhaul, is an important factor when placing a value on an airplane. The recommended TBO, less the hours currently on the engine, is the expected remaining life of the engine.

Three basic terms are normally used when referring to time on an airplane engine:

☐ *Low Time*—First ⅓ of TBO

☐ *Mid Time*—Second ⅓ of TBO

☐ *High time*—Last ⅓ of TBO

Naturally, other variables come into play when referring to TBO:

☐ Are the hours on the engine since new, remanufacture (rebuild), or overhaul?

☐ What type of flying has the engine seen? Rental and training aircraft sometimes take real beatings from uncaring/unknowing pilots.

☐ Was it flown on a regular basis? Airplanes that have not been flown on a regular basis—and maintained in a like fashion—will never reach full TBO. When an engine isn't run, acids and moisture in the oil will oxidize (rust) engine components.

The Overhauled Engine

Beware of the engine that has just a few hours on it since an overhaul. Perhaps something is not right with the overhaul, or it was a very cheap job, just to make the plane more saleable.

When it comes to overhauls, I always recommend the large shops that specialize in aircraft engine rebuilding. I'm not saying that the local FBO can't do a good job; I just feel that the large organizations specializing in this work have more experience and equipment to work with. In addition, they have excellent reputations to live up to, and will back you in the event of difficulties.

Engines are expensive to rebuild/overhaul. Here are some typical costs for a complete overhaul (based upon 1986 pricing), including installation.

Engine	Cost
O-235	$3,600 - 4,200
O-320	4,000 - 5,100
O-360	5,300 - 5,775
TSIO-360	7,600 - 8,650
O-540	6,800 - 7,500
IO-540	7,400 - 9,800

Used Airplane Prices

Used airplane prices can fluctuate to the extremes, and are dependent on more than the physical airframe and its contents. The actual selling price is the amount mutually agreed upon by the seller and buyer. This agreed-upon sum is arrived at by bargaining.

The concept of bargaining is to find the point where the seller's asking price equals the buyer's purchase price. Bargaining—the trading of price offers and counteroffers—is the norm in aircraft sales. Plan to do lots of it in purchasing an airplane. No price is ever set in concrete—not even that of a dealer!

In short, the selling price is the least amount the owner will take for his airplane, and the purchasing price is the most the buyer will pay.

Caveat Emptor

An old friend who has bought and sold airplanes for a living says: "The selling price is that asked for the one-owner, super-clean, low-time, family pride of an airplane. The purchasing price is that sum offered for the same box of rocks!''

Ever hear about the one-owner car driven by the little old lady? The used airplane salesman's equivalent is the retired airline pilot's plane. Don't you believe it; there aren't enough retired captains to own all those planes.

Lastly, the ads that read: ''Owned and maintained by licensed mechanic.'' That's real fine, but I know plumbers with leaky faucets in their own homes, and electricians who run their whole house on extension cords. Remember the old adage that ''the cobbler's children go barefoot''?

To sum it up: How badly does the seller want to sell, and how badly does the buyer want to buy?

PRICE GUIDE

The following pages are a guide to prices for many models of Piper Indians. The prices are based upon average asking/selling prices for 1986. As with most aircraft, there are no real set prices; hence, this is meant as a guide only.

The prices are for a plane with average avionics and with middle time on the engine. Middle time on the engine is defined as the middle one-third of the TBO (i.e., on a 2,000-hour engine, this would be 600 to 1300 hours SMOH).

PA-28

Model 140

Year	Asking	Year	Asking
1964	$8,000	1971	$11,000
1965	$8,000	1972	$11,000
1966	$8,500	1973	$11,500
1967	$8,500	1974	$12,000
1968	$9,000	1975	$13,000
1969	$9,500	1976	$13,500
1970	$10,000	1977	$14,000

Model 150

Year	Asking	Year	Asking
1962	$8,000	1963	$8,500

Year		Year	
1964	$9,000	1966	$9,000
1965	$9,000	1967	$9,000

Model 151

Year	Asking	Year	Asking
1974	$13,500	1976	$15,000
1975	$14,000	1977	$16,000

Model 160

Year	Asking	Year	Asking
1962	$9,000	1965	$10,500
1963	$9,500	1966	$11,000
1964	$10,000	1967	$11,750

Model 161

Year	Asking	Year	Asking
1977	$17,000	1980	$25,000
1978	$19,000	1981	$33,000
1979	$22,000	1982	$38,000

Model 180

Year	Asking	Year	Asking
1963	$11,000	1970	$17,000
1964	$12,000	1971	$17,250
1965	$13,000	1972	$17,750
1966	$14,000	1973	$18,000
1967	$14,750	1974	$20,000
1968	$15,500	1975	$21,000
1969	$16,500		

Model R180

Year	Asking	Year	Asking
1967	$18,500	1968	$19,000

| 1969 | $19,500 | 1971 | $20,000 |
| 1970 | $19,750 | | |

Model 181

Year	Asking	Year	Asking
1976	$25,000	1980	$35,000
1977	$27,000	1981	40,500
1978	$28,500	1982	$45,000
1979	$31,000	1983	$52,000

Model R200

Year	Asking	Year	Asking
1969	$20,000	1973	$23,500
1970	$21,000	1974	$25,000
1971	$22,000	1975	$26,000
1972	$22,500	1976	$26,500

Model R201

Year	Asking	
1977	$29,000	add $5,000 if turbo
1978	$33,500	add $5,000 if turbo

Model RT201

Year	Asking	
1979	$37,500	add $3,000 if turbo
1980	$43,000	add $5,000 if turbo
1981	$54,000	add $5,000 if turbo
1982	$64,000	add $7,000 if turbo
1983	n/a	
1984	n/a	
1985	n/a	

Model 235

Year	Asking	Year	Asking
1964	$15,000	1971	$21,000
1965	$16,000	1972	$22,000
1966	$17,000	1973	$23,000
1967	$17,500	1974	$24,000
1968	$18,000	1975	$25,000
1969	$19,000	1976	$28,000
1970	$20,000	1977	$32,000

Model 236

Year	Asking	Year	Asking
1979	$40,000	1982	$60,000
1980	$47,000	1983	$74,000
1981	$51,000		

PA-32

Model 260

Year	Asking	Year	Asking
1965	$20,000	1972	$26,500
1966	$21,000	1973	$27,500
1967	$21,500	1974	$29,500
1968	$22,000	1975	$31,000
1969	$23,000	1976	$33,500
1970	$23,500	1977	$36,000
1971	$24,000	1978	$38,000

Model 300

Year	Asking	Year	Asking
1966	$22,000	1970	$26,000
1967	$23,000	1971	$27,000
1968	$24,500	1972	$28,000
1969	$25,000	1973	$29,000

1974	$31,000	1977	$38,000
1975	$33,500	1978	$42,500
1976	$35,000	1979	$47,000

Model R300

Year	Asking	
1976	$42,500	
1977	$46,000	
1978	$49,500	conventional tail
1978	$45,000	T-tail
1979	$47,000	add $5,000 if turbo

Model 301

Year	Asking	
1980	$65,000	add $5,000 if turbo
1981	$72,000	add $7,000 if turbo
1982	$85,000	add $9,000 if turbo

Model R301

Year	Asking	
1980	$76,500	add $5,000 if turbo
1981	$87,000	add $7,000 if turbo
1982	$99,000	add $9,000 if turbo

PA-24

Model 180

Year	Asking	Year	Asking
1958	$13,500	1962	$18,500
1959	$15,000	1963	19,500
1960	$16,000	1964	$20,000
1961	$18,000		

Model 250

Year	Asking	Year	Asking
1958	$17,000	1962	$22,000
1959	$18,000	1963	$22,500
1960	$19,000	1964	$23,500
1961	$20,500		

Model 260

Year	Asking	
1965	$26,000	
1966	$27,500	
1967	$28,000	
1968	$29,500	
1969	$32,000	
1970	$34,000	add $5,000 if turbo
1971	$35,000	add $5,000 if turbo
1972	$37,500	add $5,000 if turbo

Model 400

Year	Asking
1964	$35,000
1965	$35,500

PA-38

Year	Asking	Year	Asking
1978	$8,000	1981	$13,000
1979	$8,500	1982	$16,500
1980	$10,700		

PAPERWORK

When you purchase an airplane, you must deal with the U.S. Government. When you deal with the government, you must fill out many forms. After all, it's paper that greases the wheels of our government, isn't it?

Title Search

The first step in the actual purchase of an airplane is to ensure that the craft has a clear title. This is done by a *title search*.

A title search is accomplished by checking the aircraft's individual records at the Mike Monroney Aeronautical Center in Oklahoma City, Oklahoma. These records include title information, chain of ownership, Major Repair/Alteration (Form 337) information, and other data pertinent to a particular airplane. The FAA files this information by "N" number.

The object of a title search is to ascertain that there are no liens or other hidden encumbrances against the ownership of the airplane. This search may be done by you, your attorney, or other representative selected by you.

Since most prospective purchasers would find it inconvenient to travel to Oklahoma City to do the search themselves, it is advisable to contract with a third party specializing in this service. For further information, contact:

AOPA
421 Aviation Way
Frederick, MD 21701
Phone: (301) 695-2000

Aircraft Title Corp.
1411 Classen Blvd.
Oklahoma City, OK 73106
Phone: (405) 685-7960

There are other organizations that provide similar services; they advertise in *Trade-A-Plane*.

In addition to title searches, AOPA offers inexpensive title insurance, which protects the owner against unrecorded liens, FAA recording mistakes, or other clouds on the title.

Documents

The following documents must be given to you with your airplane:

- ☐ Bill of Sale
- ☐ Airworthiness Certificate
- ☐ Logbooks
    - —Airframe
    - —Engine/propeller
- ☐ Equipment List (including weight and balance data)
- ☐ Flight Manual

Forms to be Completed

☐**FAA AC Form 8050-2: Bill of Sale**—Standard means of recording transfer of ownership (Fig. 10-4).

☐**FAA AC Form 8050-1: Aircraft Registration**—Application filed with the Bill of Sale, or its equivalent. The pink copy of

```
                                                              FORM APPROVED
              UNITED STATES OF AMERICA                        OMB No 2120-0029
DEPARTMENT OF TRANSPORTATION FEDERAL AVIATION ADMINISTRATION  EXP. DATE 10/31/84
              AIRCRAFT BILL OF SALE

    FOR AND IN CONSIDERATION OF $          THE
    UNDERSIGNED OWNER(S) OF THE FULL LEGAL
    AND BENEFICIAL TITLE OF THE AIRCRAFT DES-
    CRIBED AS FOLLOWS:

    UNITED STATES
    REGISTRATION NUMBER  N
    AIRCRAFT MANUFACTURER & MODEL

    AIRCRAFT SERIAL No.


    DOES THIS         DAY OF        19
        HEREBY SELL, GRANT, TRANSFER AND
        DELIVER ALL RIGHTS, TITLE, AND INTERESTS     Do Not Write In This Block
        IN AND TO SUCH AIRCRAFT UNTO:                FOR FAA USE ONLY

        NAME AND ADDRESS
        (IF INDIVIDUAL(S), GIVE LAST NAME, FIRST NAME, AND MIDDLE INITIAL.)

P
U
R
C
H
A
S
E
R

        DEALER CERTIFICATE NUMBER
AND TO          EXECUTORS, ADMINISTRATORS, AND ASSIGNS TO HAVE AND TO HOLD
SINGULARLY THE SAID AIRCRAFT FOREVER, AND WARRANTS THE TITLE THEREOF.

IN TESTIMONY WHEREOF     HAVE SET     HAND AND SEAL THIS       DAY OF    19

        NAME (S) OF SELLER       SIGNATURE (S)             TITLE
        (TYPED OR PRINTED)       (IN INK) (IF EXECUTED     (TYPED OR PRINTED)
                                 FOR CO-OWNERSHIP, ALL MUST
                                 SIGN.)

S
E
L
L
E
R

ACKNOWLEDGMENT   (NOT REQUIRED FOR PURPOSES OF FAA RECORDING: HOWEVER, MAY BE REQUIRED
BY LOCAL LAW FOR VALIDITY OF THE INSTRUMENT.)


ORIGINAL: TO FAA
AC FORM 8050-2 (8-76) (0082-629-0002)
```

Fig. 10-4. FAA AC Form 8050-2.

the registration is retained by you and will remain in the airplane until the Certificate of Aircraft Registration (FAA AC Form 8050-3) is issued by the FAA (Figs. 10-5, 10-6).

☐ **FAA AC Form 8050-41: Release of Lien**—Must be filed by the seller if a lien is recorded.

☐ **FAA AC Form 8050-64: Assignment of Special Registration Number**—for ''vanity'' registration.

☐ **FCC Form 404: Federal Communications Commission Application for Aircraft Radio Station License**—Must be completed if you have any radio equipment on board. The tear-off section will remain in your airplane as temporary authorization until the new license is sent to you (Fig. 10-7).

Most forms sent to the FAA or FCC will result in the issuance of a document to you. Be patient; it all takes time.

Paperwork Assistance

Although not complicated, there are many forms to be completed when purchasing an airplane, and you may wish to seek assistance in filling them out. You can check with your FBO, or call upon another party, such as the AOPA.

The AOPA, for a small fee, will provide closing services via telephone, and prepare/file the necessary forms to complete the transaction. This is particularly nice if the parties involved in the transaction are spread all over the country, as would be the case if you are purchasing an airplane ''sight unseen.''

Another source of assistance in completing the necessary paperwork is your bank. This is particularily true if the bank has a vested interest in your airplane (i.e., they hold the note!).

INSURANCE

Few people can afford to take big risks. Insure your airplane from the moment you sign on the dotted line.

Basically, there are two types of insurance you will be concerned with.

Liability insurance protects you, or your heirs, in instances of claims against you, or your estate, resulting from your operation of an airplane (i.e., bodily injury or property damage, death). In this age of litigation, you *will be* sued if anyone is injured or killed while riding in your airplane, or struck by it on the ground.

189

UNITED STATES OF AMERICA DEPARTMENT OF TRANSPORTATION
FEDERAL AVIATION ADMINISTRATION-MIKE MONRONEY AERONAUTICAL CENTER
AIRCRAFT REGISTRATION APPLICATION

CERT. ISSUE DATE

UNITED STATES
REGISTRATION NUMBER **N**

AIRCRAFT MANUFACTURER & MODEL

AIRCRAFT SERIAL No.

FOR FAA USE ONLY

TYPE OF REGISTRATION (Check one box)

☐ 1. Individual ☐ 2. Partnership ☐ 3. Corporation ☐ 4. Co-owner ☐ 5. Gov't ☐ 8. Foreign-owned Corporation

NAME OF APPLICANT (Person(s) shown on evidence of ownership. If individual, give last name, first name, and middle initial.)

TELEPHONE NUMBER: (   )   –

ADDRESS (Permanent mailing address for first applicant listed.)

Number and street: _____

Rural Route: _____   P.O. Box: _____

CITY	STATE	ZIP CODE

☐ **CHECK HERE IF YOU ARE ONLY REPORTING A CHANGE OF ADDRESS**

ATTENTION! Read the following statement before signing this application.

A false or dishonest answer to any question in this application may be grounds for punishment by fine and / or imprisonment (U.S. Code, Title 18, Sec. 1001).

CERTIFICATION

I/WE CERTIFY:

(1) That the above aircraft is owned by the undersigned applicant, who is a citizen (including corporations) of the United States.

(For voting trust, give name of trustee: _____). or:

CHECK ONE AS APPROPRIATE:

a. ☐ A resident alien, with alien registration (Form 1-151 or Form 1-551) No. _____

b. ☐ A foreign-owned corporation organized and doing business under the laws of (state or possession) _____ , and said aircraft is based and primarily used in the United States. Records of flight hours are available for inspection at _____

(2) That the aircraft is not registered under the laws of any foreign country; and
(3) That legal evidence of ownership is attached or has been filed with the Federal Aviation Administration.

NOTE: If executed for co-ownership all applicants must sign. Use reverse side if necessary.

TYPE OR PRINT NAME BELOW SIGNATURE

EACH PART OF THIS APPLICATION MUST BE SIGNED IN INK.	SIGNATURE	TITLE	DATE
	SIGNATURE	TITLE	DATE
	SIGNATURE	TITLE	DATE

NOTE: Pending receipt of the Certificate of Aircraft Registration, the aircraft may be operated for a period not in excess of 90 days, during which time the PINK copy of this application must be carried in the aircraft.

AC FORM 8050-1 (1-83) (0052-00-628-9005)

Fig. 10-5. FAA AC Form 8050-1.

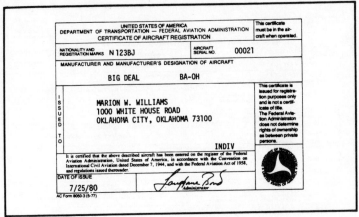

Fig. 10-6. FAA AC Form 8050-3, the final aircraft registration that you will receive and display in your airplane.

Hull insurance protects your investment from loss caused by the elements of nature, fire, theft, vandalism, or accident. There are limited-coverage policies available that provide for losses to the airplane while on the ground, but not while in the air. You can save money here; however, discussion of coverages available is best left between you and your insurance agent. Your lending institution will require hull insurance for their protection.

A check of any of the various aviation publications will produce telephone numbers for several aviation underwriters. The larger insurance companies have "800" toll-free telephone numbers. Call them; it's free! Call *all* of them, as services, coverage, and rates do differ.

Don't buy a policy that has complicated exclusions, or other specific rules involving maximum pre-set values for replacement parts or payment of losses. Purchase a policy that you can read and understand, one that is written in "lay English."

Something else to consider is your personal health and life insurance coverage. Be sure you are covered while flying a "small" airplane. Some policies do not cover pilots or passengers, and in the event of injury or death there might be no payoff.

SAFEKEEPING AN AIRPLANE

There is no way to make your airplane theftproof; however, it can be made less attractive to the thief. "Less attractive" means more difficult to steal or break into.

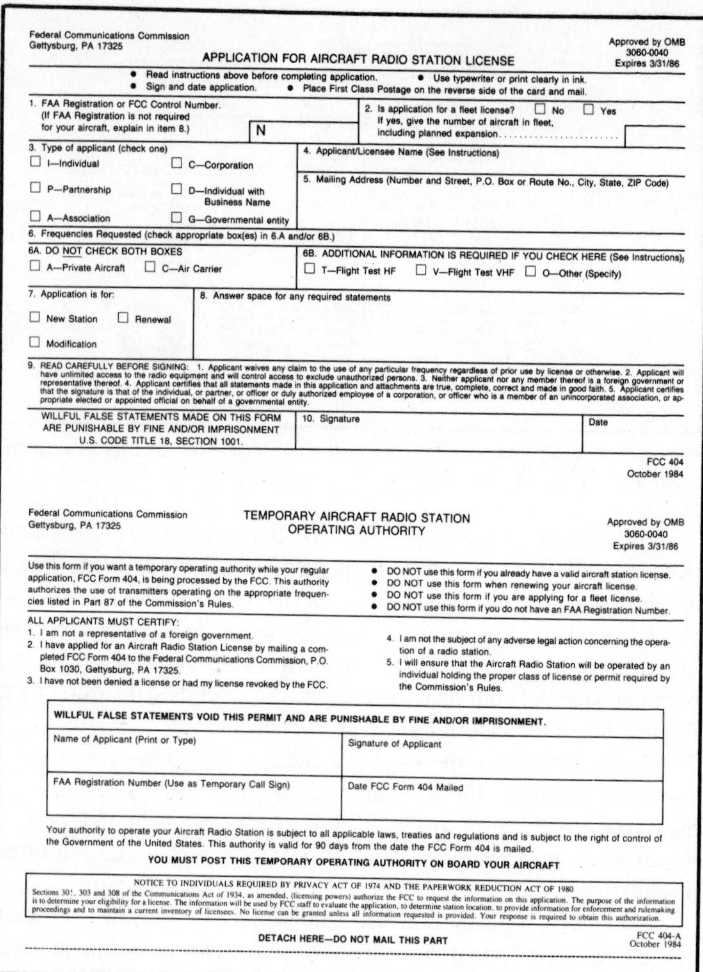

Federal Communications Commission
Gettysburg, PA 17325

Approved by OMB 3060-0040 Expires 3/31/86

APPLICATION FOR AIRCRAFT RADIO STATION LICENSE

- Read instructions above before completing application.
- Use typewriter or print clearly in ink.
- Sign and date application.
- Place First Class Postage on the reverse side of the card and mail.

1. FAA Registration or FCC Control Number. (If FAA Registration is not required for your aircraft, explain in item 8.) N

2. Is application for a fleet license? ☐ No ☐ Yes
If yes, give the number of aircraft in fleet, including planned expansion........................

3. Type of applicant (check one)
☐ I—Individual ☐ C—Corporation
☐ P—Partnership ☐ D—Individual with Business Name
☐ A—Association ☐ G—Governmental entity

4. Applicant/Licensee Name (See Instructions)

5. Mailing Address (Number and Street, P.O. Box or Route No., City, State, ZIP Code)

6. Frequencies Requested (check appropriate box(es) in 6.A and/or 6B.)

6A. DO NOT CHECK BOTH BOXES
☐ A—Private Aircraft ☐ C—Air Carrier

6B. ADDITIONAL INFORMATION IS REQUIRED IF YOU CHECK HERE (See Instructions)
☐ T—Flight Test HF ☐ V—Flight Test VHF ☐ O—Other (Specify)

7. Application is for:
☐ New Station ☐ Renewal
☐ Modification

8. Answer space for any required statements

9. READ CAREFULLY BEFORE SIGNING: 1. Applicant waives any claim to the use of any particular frequency regardless of prior use by license or otherwise. 2. Applicant will have unlimited access to the radio equipment and will control access to exclude unauthorized persons. 3. Neither applicant nor any member thereof is a foreign government or representative thereof. 4. Applicant certifies that all statements made in this application and attachments are true, complete, correct and made in good faith. 5. Applicant certifies that the signature is that of the individual, or partner, or officer or duly authorized employee of a corporation, or officer who is a member of an unincorporated association, or appropriate elected or appointed official on behalf of a governmental entity.

WILLFUL FALSE STATEMENTS MADE ON THIS FORM ARE PUNISHABLE BY FINE AND/OR IMPRISONMENT U.S. CODE TITLE 18, SECTION 1001.

10. Signature

Date

FCC 404
October 1984

Federal Communications Commission
Gettysburg, PA 17325

TEMPORARY AIRCRAFT RADIO STATION
OPERATING AUTHORITY

Approved by OMB 3060-0040 Expires 3/31/86

Use this form if you want a temporary operating authority while your regular application, FCC Form 404, is being processed by the FCC. This authority authorizes the use of transmitters operating on the appropriate frequencies listed in Part 87 of the Commission's Rules.

- DO NOT use this form if you already have a valid aircraft station license.
- DO NOT use this form when renewing your aircraft license.
- DO NOT use this form if you are applying for a fleet license.
- DO NOT use this form if you do not have an FAA Registration Number.

ALL APPLICANTS MUST CERTIFY:

1. I am not a representative of a foreign government.
2. I have applied for an Aircraft Radio Station License by mailing a completed FCC Form 404 to the Federal Communications Commission, P.O. Box 1030, Gettysburg, PA 17325.
3. I have not been denied a license or had my license revoked by the FCC.
4. I am not the subject of any adverse legal action concerning the operation of a radio station.
5. I will ensure that the Aircraft Radio Station will be operated by an individual holding the proper class of license or permit required by the Commission's Rules.

WILLFUL FALSE STATEMENTS VOID THIS PERMIT AND ARE PUNISHABLE BY FINE AND/OR IMPRISONMENT.

Name of Applicant (Print or Type)	Signature of Applicant
FAA Registration Number (Use as Temporary Call Sign)	Date FCC Form 404 Mailed

Your authority to operate your Aircraft Radio Station is subject to all applicable laws, treaties and regulations and is subject to the right of control of the Government of the United States. This authority is valid for 90 days from the date the FCC Form 404 is mailed.

YOU MUST POST THIS TEMPORARY OPERATING AUTHORITY ON BOARD YOUR AIRCRAFT

NOTICE TO INDIVIDUALS REQUIRED BY PRIVACY ACT OF 1974 AND THE PAPERWORK REDUCTION ACT OF 1980
Sections 301, 303 and 308 of the Communications Act of 1934, as amended, (licensing powers) authorize the FCC to request the information on this application. The purpose of the information is to determine your eligibility for a license. The information will be used by FCC staff to evaluate the application, to determine station location, to provide information for enforcement and rulemaking proceedings and to maintain a current inventory of licensees. No license can be granted unless all information requested is provided. Your response is required to obtain this authorization.

DETACH HERE—DO NOT MAIL THIS PART

FCC 404-A
October 1984

Fig. 10-7. Application portion of the FCC Form 404 (at top), which you will mail to the FCC. On the bottom is the portion of the form you will retain in your airplane as temporary operating authority.

The thief doesn't want to spend large amounts of time in entering an airplane. He wants to get in and go. If you can delay him, you might discourage him.

There are several methods of delaying the thief; all involve locks of one type or another. Store the airplane in a locked hangar. Use cut-resistant chain and locks for the tiedowns. Install a throttle lock, an excellent device for discouraging the would-be thief, available from:

Spiser Aircraft Co., Inc.
Municipal Airport
Clay Center, KS 67432
Phone: (913) 632-3217

Unfortunately, there is now another type of criminal loose in America. This type will attempt to steal your property, and if unable to do so, will destroy it. His reasoning: If he can't have it, neither can you.

This criminal element is primarily found in urban areas breaking into Mercedes, BMW, and Volvo automobiles, but this disease has spread to some "close-in" airports during recent years.

Reporting a Stolen Airplane

What would you do if you drove out to the airport and your airplane had been stolen, or it had been broken into and some of your avionics were missing?

Naturally, you would notify your local police. They will come to the airport—maybe—and make a report of the theft, and possibly process the crime scene by attempting to lift fingerprints. The latter will be accomplished only if there is a "chance for prints." There is no chance for prints if it has rained since the break-in, or if the scene has been contaminated by you or others touching the airplane.

Don't expect the police to do very much about your loss. The reports will be filed and entries of registration and serial numbers will be made into the National Crime Information Center (NCIC) computer. This will give a chance of recovery in the event that another police department on the other side of the country comes into contact with the stolen items.

Notify the FAA. They will issue a nationwide stolen-aircraft alert. If the registration numbers are not changed, and a controller is sharp, you have a chance of recovery.

Notify the IATB (International Aviation Theft Bureau) at (301) 695-2022 (Telex: 89-3445). The IATB is a part of the AOPA operation.

Notify your insurance company of the loss, and be ready to supply them with copies of all police reports, purchase receipts, etc.

Chapter 11

Caring for Your Airplane

There are four basic areas of care the airplane owner needs to be concerned with. These are:

☐ Proper ground handling
☐ Effective storage
☐ Cleaning and protecting
☐ Preventive maintenance (Chapter 12 of this book is devoted entirely to preventive maintenance)

All four areas provide the owner with methods of becoming intimately acquainted with his airplane. Additionally, by properly caring for the craft, you will protect your investment in the airplane, increase flying safety, and save a considerable amount of money.

GROUND HANDLING

Proper ground handling of an airplane (towing, parking, and mooring), especially when done by hand, is extremely important. Sloppy ground handling can inflict major structural damage that could cost thousands of dollars to repair.

Towing

Moving an airplane is normally accomplished by use of a nosewheel steering bar (towbar). This bar fastens to the nosewheel and permits pulling, pushing, and turning of the airplane. Do not tow the airplane with the control locks in place.

Also, *never* use the propeller as a push point! There are two reasons for this. Pushing on the propeller could easily turn the engine slightly, causing it to start up—resulting in severe injury or death.

Pushing on the propeller also causes flexing (bending), which can lead to catastrophic blade failure. Blade failure means complete separation caused by metal fatigue due to the stress of bending.

I see (as will you) airplanes pushed and pulled by their propellers all the time. *Don't do it.*

Parking

Parking is generally considered the *temporary* stopping of the aircraft. Loading, refueling, and nature calls are all examples of brief parking periods. Parking never means leaving the airplane sitting, not tied down, for more than a few moments.

All the Piper Indian airplanes have parking brakes. These brakes should be set whenever the airplane is parked, with the following exceptions:

☐ Never apply the parking brakes when the brake system is overheated.

☐ Never set the parking brakes when moisture is present at below-freezing or near-freezing temperatures.

Mooring

To fully protect an airplane when left unattended, it must be moored (tied down) properly. An airplane tied down cannot be moved or damaged by the wind (Fig. 11-1).

An airplane is tied down at three points: each side and the tail. By properly securing each of these points, the aircraft is unable to move freely in the wind.

Routinely, aircraft tiedown procedures call for facing the airplane into the wind, then mooring it. In most cases this is not practical, as fixed tiedown systems have been placed by the FBO.

Typical tiedown systems are either individual anchors with eyes for rope connections, or parallel wire ropes with loops for rope connections. Chains or small cables may be used in place of rope.

Never depend on hand-driven stakes for tiedown purposes (Fig. 11-2). They are too easy to pull out, allowing your airplane to move and possibly become damaged.

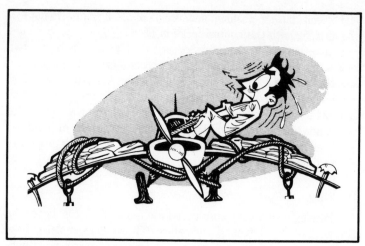

Fig. 11-1. Moor your airplane securely, or else it might not be there when you come looking for it! (courtesy FAA)

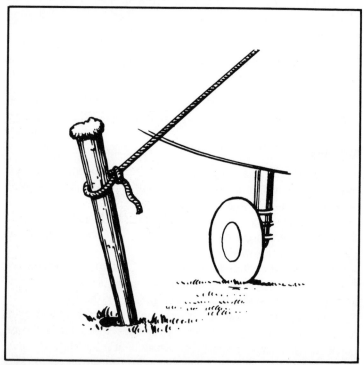

Fig. 11-2. Wooden stakes can work themselves out of the ground; then your airplane is free of its "surly bonds" and may do whatever it pleases. (courtesy FAA).

196

Tiedown tethers should be fastened to the wing tiedown rings and to the tail skid. Additionally, a fourth line may be attached to the nosewheel forks.

Ropes used for tiedowns should be capable of withstanding a pull in excess of 3,000 pounds. There are two basic types of rope used for tiedowns: manila and synthetic fiber.

Manila ropes are susceptible to rotting and attack by fungus. They lose strength with age, and often shrink when wet.

Dacron and nylon ropes are preferred over manila due to their superior durability and strength. However, the synthetic ropes must be tied very carefully, as knots tend to work out easily. The recommended knots for use with synthetic ropes are the square or bowline knots.

Ropes are cheap when compared to the cost of an airplane. They should be replaced annually, due to the weathering that constantly takes place. Manila rots from dampness, and the synthetics are attacked by ultraviolet rays from the sun. A long-lived replacement for rope is chain.

Always install control locks when mooring an airplane. These will stop the surfaces from banging in the wind, thereby preventing costly damage. Additional protection may sometimes be necessary to prevent wind damage to the control surfaces—in particular, to the rudder (Fig. 11-3).

Fig. 11-3. Control surface locks can be installed as extra protection. Be sure to put streamers on the locks to remind you to remove them before flight.

Install a pitot tube cover, and in the case of the PA-28 retractables, cover the gear actuator device on the left side of the cabin. These covers keep out the mud daubers and other insects that could plug these devices.

While plugging holes, be sure to plug the engine air intakes. This will keep out birds (Figs. 11-4, 11-5).

Heat and Sunlight Protection

The interior temperature of a parked aircraft can reach as much as 185°F. This heat buildup will not only damage avionics, but will cause problems with instrument panels, upholstery, and various other "plastic things." High temperatures are extremely destructive.

A quick look around the local airport will show four methods of heat protection for the airplane cabin:

☐ None at all.
☐ A chart or towel on the panel.
☐ An interior reflective cover.
☐ An exterior protective cover.

The first method—no protection—is noticed most often on old-

Fig. 11-4. These air intake plugs keep out the birds. Notice the rope wrapped around the prop. It fouls the prop, and will prevent you from flying with the plugs in place.

Fig. 11-5. The installation of hardware cloth (screening) keeps the birds out of this plane's engine compartment. It is a permanent fix for the problem, and requires no removal prior to flying.

er, already sad-looking airplanes. There is certainly no ownership pride here—and the situation will only get worse.

Some owners of older planes recognize the need for protection from the sun's rays, so they lay a towel or chart over the top of the instrument panel. This is merely an exercise in futility, as no heat-buildup protection is afforded this way. The only effective use for this is when parking for a short period of time, such as refueling (Fig. 11-6).

An interior reflective cover protects the cabin of the aircraft by reflecting away the sun's rays. A metallic-type reflective surface is on one side of most interior covers. These covers are normally installed by means of Velcro fasteners. Interior reflective covers are available from many sources and are advertised in all the aviation periodicals (Fig. 11-7).

Exterior airplane covers provide similar protection for the interior of the aircraft, yet give additional exterior protection by covering the windshield, roof, and fresh air vents (Fig. 11-8).

STORAGE

There is much more to proper storage of an aircraft than the mere act of hangaring or parking.

Fig. 11-6. Covering the instrument panel like this is good while fueling up, but will not give long-term protection.

So often, I watch airplanes sit on the field with no use or attention for months on end. This is particularly true from November until May in many areas of the country. Not one time does the owner come by to check on his plane. He tied it down after his last flight and left. Later—perhaps several *months* later—he reappears, normally on a warm clear Saturday morning with the glint of "let's fly" in his eyes. Without a care, he walks around his plane once,

Fig. 11-7. The metallic reflective cover seen through the windows of this PA-38 keeps much of the heat out of the cockpit.

Fig. 11-8. This external cover not only keeps heat out of the cockpit, it keeps the windows clean.

unties it, climbs in, and starts it. I could go on, but I'm sure by now you have the picture.

Unfortunately, a large number of airplane owners pay little heed to the proper preserving of their airplanes during periods of non-use. The results from improper care are unsafe airplanes and very high maintenance bills. The example above can shorten engine life by 50 to 75 percent, to say nothing of the airframe problems that will crop up after a few seasons of neglect.

For complete information regarding airframe and engine storage, consult the aircraft's manuals. Improper storage is one of the most important reasons to be so *very* careful when purchasing a used airplane.

CLEANING THE AIRPLANE

Airplane ownership is a source of pride. To display this pride, you should have a clean airplane. But pride alone is not the real reason for keeping the appearance of an airplane in top shape. A clean airplane retains its value, indicates that the owner cares for his airplane, and forces close inspection of the aircraft during cleaning.

Exterior Care

Complete washing with automotive-type cleaners will produce good results, and the materials used will be much cheaper than so-called "aircraft cleaners." Automotive protective coatings—formerly called *wax*—will protect painted and unpainted surfaces. The new "space age" silicone preparations are very easy to apply, don't whiten rubber components, and will protect your airplane's finish for many months. Just remember, there are a lot of surfaces on an airplane . . . many square feet. So use the best products available, unless you like to make a career of airplane polishing.

When cleaning windshields and windows, use clean water and very mild soap. Wash with a soft cloth; never scrub! Never allow gasoline, alcohol, acetone, paint thinner, or spray window cleaners to come into contact with the plastic windows. Spray window cleaners will cloud the surface, rendering the window useless.

Cleaning Materials

The following is a sampling of some of the available products for cleaning and preserving your aircraft's exterior. Most should be familiar to the automobile owner.

Spray engine cleaner is used for degreasing the engine area and front strut. It dissolves grease and can be washed off with water. When using the cleaner inside the engine compartment, you must cover the magnetos and alternator with plastic bags to keep out the cleaner and rinse-water. Engine cleaners are also good for cleaning the belly of an oil-stained airplane.

Rubbing compound is used to clean away stains caused by engine exhaust. Rubbing compounds are available in several strengths (abrasiveness); use the mildest. Avoid being over-zealous in the application of rubbing compound, as you can easily remove paint by overdoing it.

Novus Polish #2 is for windshield maintenance. Mildly abrasive, it will polish out scratches and pitmarks that collect on Plexiglas. In effect, it is a very mild polishing compound.

Bugaway is a windshield cleaning solvent developed especially for aircraft use. This spray solution dissolves insect splatters and bird droppings instantly. Just spray it on, wait 10 seconds, then wipe clean with a soft cloth. Bugaway will not harm aluminum or painted surfaces. Bugaway and Novus Polish #2 are available from:

Connecticut Aviation Products, Inc.
P.O. Box 12
East Glastonbury, CT 06025

Interior Care

The interior of the airplane is seen and judged by all, including the pilot and his passengers. Unfortunately, keeping the inside of an airplane clean can be a major problem. As in exterior care, I recommend using automobile cleaning products.

In addition to the more normal cleaning items used on the family automobile, there are other grocery-store-type products that will greatly assist you in keeping the interior of your airplane clean and shining.

Spray furniture wax can be used on most hard surfaces (including the windshield) and vinyl surfaces (seats, dash panel and doors). The lemon smell is nice, too.

Heavy-duty spray cleaner is good for the hard-to-remove grease and dirt smudges. Keep it away from the windshield, instruments, and painted surfaces.

Window cleaner is an excellent product for small cleanup jobs. However, *never use it on the windshield or other windows!* The ammonia found in most window cleaners will cause a characteristic clouding of plastic windows, spoiling the clean "see-through" qualities. In time, the clouded window will have to be replaced.

Armor-All is good as a final coat on the dash panel, kick panels, vinyl seats, etc. It makes vinyl look and smell new.

Scotch Guard is a spray-on product for seats, carpets, and other cloth areas. It will allow quick mop-up of small spills and prevent most liquids from soaking into upholstery.

WD40 is a general spray lubricant used to stop squeaks and ease movement. It's good on cables, controls, seat runners, door hinges, and latches. *Keep it off the windows.*

A shop vacuum or household cannister vacuum is very useful for cleaning the interior of an airplane. No amount of hand sweeping can remove the debris a vacuum cleaner can pick up.

Chapter 12
Preventive Maintenance

In the preceding chapter we discussed general airplane care. In this chapter we shall encounter mechanical work the owner can do without having a mechanic's license. This does not, however, eliminate the need for a licensed mechanic, as airplanes are replete with complex and sensitive workings that require expertise (and an FAA license) to properly service and repair. When a question arises about maintenance, *consult a competent licensed mechanic*. This consultation will probably cost a few dollars, but they will be dollars well spent.

THE FAA SAYS

In rare instances, local FBOs have made attempts to stop airplane owners from servicing and maintaining their own airplanes. No doubt there is a profit-motive problem; however, the FAA advises in Advisory Circular 150/5190-2A that such restrictions are illegal at airports that have received Federal development funds:

> d. *Restrictions on Self-Service.* Any unreasonable restriction imposed on the owners and operators of aircraft regarding the servicing of their own aircraft and equipment may be considered as a violation of agency policy. The owner of an aircraft should be permitted to fuel, wash, repair, paint, and otherwise take care of his own aircraft, provided there is no attempt to perform such services for others.

Restrictions which have the effect of diverting activity of this type to a commercial enterprise amount to an exclusive right contrary to law.

By issuing this circular, the FAA has allowed the owner of an aircraft to save his hard-earned dollars, and to become very familiar with his airplane, the latter no doubt contributing to safety.

The Federal Aviation Regulations (FARs) specify that preventive maintenance may be performed by pilots/owners of airplanes not utilized in commercial service.

FAR Part 43 defines *preventive maintenance* as "simple or minor preservation operations and the replacement of small standard parts not involving complex assembly operations."

Appendix A of FAR Part 43.13 lists 28 preventive maintenance items. This means that *only* those functions listed are considered preventive maintenance. If a function is not listed, it is *not* considered preventive maintenance. Further, because of differences in aircraft, a function may be preventive maintenance on one model of airplane, *yet not on another*. Remember the phrase "not involving complex assembly operations."

Owners/pilots must exercise good judgment in determining if a function listed is in fact "preventive maintenance" for their particular aircraft.

The FARs also require that all preventive maintenance work must be done in such a manner, and by use of materials of such quality, that the airframe, engine, propeller, or assembly worked on will be at least equal to its original condition.

GETTING ASSISTANCE

Before embarking on a preventive maintenance program, it is absolutely necessary that you have a complete service manual available for reference (Fig. 12-1). I recommend that you *purchase* this item, not borrow it. Current model service manuals are available from Piper, and may be ordered through your dealer. Out-of-print service manuals, such as for the older PA-28s, are available from some suppliers that advertise in *Trade-A-Plane*. Three such sources I recommend are:

J.W. Duff Aircraft Co., Inc.
8131 E. 40th Ave.
Denver, CO 80207
Phone: (303) 399-6010

Univair Aircraft Corp.
2500 Himalaya Rd.
Aurora, CO 80011
Phone: (303) 364-7661

ESSCO
Akron Airport
Akron, OH 44306
Phone (800) 228-6200
(216) 733-6241

In addition to having a service manual, I strongly advise that before you undertake any of these allowable preventive maintenance procedures, you discuss your plans with a licensed mechanic. The instructions/advice you receive from him may help you avoid making costly mistakes. You will have to pay the mechanic for his time, but after all, his time is his money.

Properly performed preventive maintenance gives the pilot/owner a better understanding of his airplane, affords substantial maintenance savings, and gives a feeling of accomplishment.

PREVENTIVE MAINTENANCE ITEMS

The following procedures are considered preventive maintenance on the Piper Indians.

Lubrication

Lubrication of the airplane includes engine oil changes and greasing (oiling) of airframe parts.

Engine Lubrication: To drain the engine oil, remove the engine cowl and open the oil drain valve on the underside of the engine. Be sure to drain the oil into a suitable container and dispose of the old oil in an environmentally acceptable manner. It is easier to drain the oil if the engine is first warmed to operating temperature, but be careful with hot oil.

The oil screen must be removed and inspected for trapped metal particles. Also, sludge should be cleaned away. The exact location of the oil screen will vary from one type of engine to another, so refer to your trusty service manual.

A full-flow oil filter must be replaced with a new unit before new engine oil is added. After the filter is removed, it is good practice to cut the filter can open and inspect the internal elements for metal particles. Metal particles found on the oil screen or inside the filter element *must* be brought to your mechanic's attention.

Fill the engine by pouring the proper type and weight of

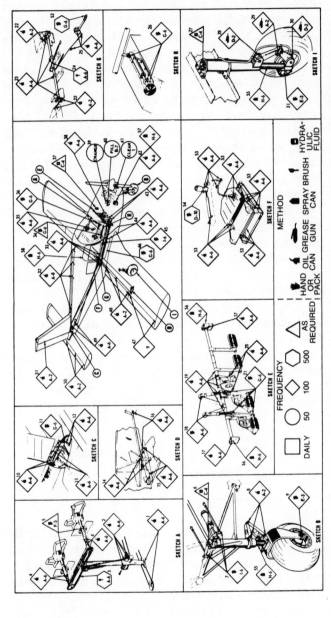

Fig. 12-1. This diagram, from the maintenance manual, indicates the particulars of lubrication for the PA-28-140: where, how, and what type of lubrication will be used. This chart is a perfect example of why you should own a complete maintenance manual. (courtesy Piper Aircraft)

207

oil into the oil filler tube. Check the dipstick for proper level.

Oil Change Frequency. When using higher-leaded fuels (100LL), the recommended oil change period of 50 hours must not be exceeded. It is not uncommon for oil changes to be made on a 25-hour basis. This is particularly true for older engines designed for 80/87 fuel now using 100LL.

Many of the engine deposits formed by the use of the higher-leaded 100LL fuel are in suspension within the engine oil, yet are not removed by a full-flow filter. When large amounts of these contaminants in the oil reach a high-temperature area of the engine, they can be "baked out." This baking out—and the subsequent deposits left in such areas as the exhaust valve guides—can cause sticking valves. If sticking valves are noted, all valve guides should be reamed and the frequency of oil changes increased.

Engine Oil Analysis. An engine oil analysis program calls for the submission of oil samples to a laboratory for metal content analysis. Generally, the samples are submitted at specific intervals during an engine's life. For engines found in general aviation aircraft, this would be at each oil change, or at least every other change.

The purpose of an oil analysis program is to detect problems before they cause excessive damage and costly repairs. These problems cause the oil to have an abnormally high content of one or more of the nine elements (mostly metals) tested for. These elements are:

☐ Tin   ☐ Silver

☐ Nickel   ☐ Chrome

☐ Iron   ☐ Aluminum

☐ Magnesium ☐ Silicon (dirt)

☐ Copper

Not only can impending failures be detected, but the suspect area of the engine can generally be identified. Identification is based upon the different metals utilized for rings, valves, valve guides, bearings, etc. Each will leave certain elements as it wears.

Obtaining an oil analysis is very simple: Place a small sample of the freshly drained engine oil in a clean container and include certain data about your engine: serial number, model number, hours since new/rebuild, aircraft registration number, your name and address. When you obtain the results from the testing facility, enter the information in the engine log for future reference.

A testing kit (Fig. 12-2), including sample bottle, blank engine information sheet, instructions, and a mailer, is available from:

> Engine Oil Analysis
> 7820 South 70th East Ave.
> Tulsa, OK 74133
> Phone: (918) 665-6464

A small fee is charged for the analysis, and should be remitted when the sample is sent to the testing facility. Call for the current price.

Airframe Lubrication: The object of proper lubrication is to reduce wear. Use only the recommended lubricants. Clean and check each lubrication point before applying fresh lubricant. Fresh engine oil is an allowable substitute for general-purpose lubrication oil.

The lubrication chart in your service manual depicts each

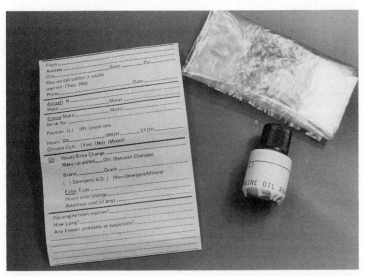

Fig. 12-2. Complete engine oil sample kit, available from Engine Oil Analysis.

part needing lubrication and indicates the time interval between lubrications and the method of lubrication.

Fluid Levels

A good time to check fluid levels and the battery is while lubricating the plane. Unless otherwise indicated in the service manual:

☐ The fluid level in the brake cylinder reservoir should be checked every 50 hours. To replenish, fill to the level marked on the reservoir with MIL-H-5606 fluid.

☐ The hydraulic fluid reservoir (retractables only) should be checked and filled with MIL-H-5606 fluid to the proper level every 50 hours of operation.

☐ The battery should be serviced by adding distilled water to maintain the electrolyte level. This level should be even with the horizontal baffle plate at the bottom of the filler holes. Be sure to flush the area with plenty of clean water after refilling to wash away any spilled "battery acid." Battery acid is very corrosive, and can do considerable damage in a relatively short period of time.

Spark Plugs

Spark plugs serve the primary function of igniting the fuel-air mixture in the combustion chamber of an engine. To efficiently accomplish this, spark plugs must:

☐ be capable of conducting high-voltage charges.

☐ be durable and long-lived.

☐ dissipate the heat received during combustion.

There are two types of spark plugs available, the *massive electrode* and the *fine wire electrode*. The massive electrode is the commonly utilized spark plug. The fine wire electrode utilizes precious metals that enhance the self-cleaning properties of this type of spark plug.

Spark plugs must be removed periodically and inspected, cleaned, or replaced. This procedure has become very important to the owners of older airplanes that were designed to use 80/87 fuel.

Spark Plug Removal is as follows:

1. Loosen the lead terminal unit at the top of the spark plug with an open-end wrench.

2. Hold the terminal lead elbow with one hand to avoid twisting the lead and unscrew the nut by hand.

3. Pull the lead connector straight out (apply no side pressure).

4. Inspect the terminal connector, and replace it if cracked or deteriorated.

5. Remove the spark plug with a deep socket wrench designed for the job. Identify each lug as to cylinder number and position (top or bottom) as it is removed from the engine. A spark plug rack is helpful at this point, but an egg box will work just as well.

Inspection of Spark Plugs: Write to any of the spark plug manufacturers and ask for a reference chart that will aid you in plug inspections.

A normal plug will have brownish-gray deposits on it, and slight electrode wear. Such a plug may be cleaned, regapped, and reinstalled.

A lead-fouled plug will have a tan or brown buildup on the firing end caused by the tetraethyl lead found in avgas. Lead fouling is a constant problem in aircraft using 100LL as a replacement for 80/87 fuel. Proper engine operation, as noted in Chapter 13, will reduce this type of fouling, or you can try unleaded automobile gasoline.

Spark plugs that have cracked, broken, or loose insulators or excessively worn electrodes must be replaced.

Spark Plug Installation is as follows:

1. Assure that the cylinder-head threads and spark plug threads are clean.

2. After regapping, install a new gasket, and seat it.

3. Sparingly apply an anti-seize compound to the top threads.

4. Thread the plug into the engine by hand. *Caution*: Avoid cocking or tilting the plug when screwing it in, as thread damage could result.

5. Tighten the plug with a torque wrench to 30 foot-pounds. *Note*: If a spark plug is dropped on a hard surface, throw it away.

6. Clean the ceramic terminal connector sleeves with solvent such as alcohol, acetone, or white gas.

7. Insert the terminal sleeve straight into the plug connector, and start the terminal nut by hand.

8. Tighten the terminal nut with a wrench.

Landing Gear Service

The servicing of landing gear includes jacking of the aircraft, removal of wheels, wheel disassembly, cleaning and greasing wheel bearings, tire replacement, and adding oil, air, or both to the landing gear struts.

Jacking: Place jacks under the jack pads along the front wing spar. Attach a tail support to the tail skid, and weight the support down with about 250 pounds of ballast (sand). Raise the jacks until the wheels are clear of the ground. Some late models have jack pads near the nose wheel, precluding the need of a tail support. If you are unsure of the location of the jack pads, refer to your service manual.

Nosewheel Removal: Remove the wheel fairing if installed. Remove the nut and washer from one end of the axle rod, and pull same out from the wheel and fork. Carefully remove the axle tube from the center of the wheel assembly. Remove the spacer tubes and wheel from the fork.

Main Wheel Removal: Remove the two cap bolts that hold the brake cylinder housing and brake backplate together, and remove the backplate from between the brake disc and wheel. Remove the axle dust cover, take out the cotter pin, and remove the wheel nut. Slide the wheel from the axle.

Wheel Disassembly: Deflate the tire completely, then remove the through bolts. Split the wheel by pulling apart the two halves. *Caution*: Deflate the tire *completely* before attempting wheel disassembly, or *serious injury* could result.

Inspection of the Wheels: Visually inspect the wheel for cracks, distortion, excessive wear, and defects. Check the through wheel bolts for signs of wear. Inspect the grease seals and bearings. Replace faulty parts and damaged wheels.

Tire Service: Inspect the tire casing for cuts, internal bruises, deterioration, and excessive wear. If the tire is unserviceable, replace it. Replace the valve stem at the same time.

Wheel Bearing Service: Remove the bearing cap, seal retainers, and bearing cones. Clean the bearings and wheel hub thoroughly with solvent. Repack the bearings with grease, being sure that grease is forced between the rollers in the retaining ring. Do not put grease in the wheel hub.

Wheel Reassembly: Place the tire on the wheel half with the valve stem, and press the two wheel halves together. Install the through wheel bolts (with washers and nuts) and torque to 90 inch-pounds. Inflate the tire to specified pressure (generally 24 psi).

Nosewheel Installation: Place the spacer tubes on each side of the wheel, and align the wheel on the fork while sliding the axle tube through the spacer tubes and wheel assembly. Reinstall the axle plugs and rod with washer and nut. Torque axle nut to 45 inch-pounds. Reinstall the fairing.

Main Wheel Installation: Slide the wheel on the axle, and tighten the axle nut until there is no side play, yet the wheel freely rotates. Safety the nut with a flathead pin, washer, and cotter pin. Reinstall the axle dust cover. Reassemble the brake cylinder housing and backplate units.

Landing Gear Struts: With the airplane on jacks, depress the air valve core on the top of the strut assembly. When the air pressure has dropped to zero, remove the filler plug from the top of the nose gear strut (top inboard side of the main gear struts), then compress the strut fully. Add fluid (MIL-H-5606) through the filler opening until the level reaches the bottom of the filler plug hole. Install the filler plug finger tight and exercise the strut two or three times, then remove the filler plug again. Add more fluid if needed, tightly install the filler plug, and exercise the strut two or three more times. Then, using a strut pump, inflate until the correct amount of strut piston is exposed.

Surface Repairs

Fairings and cowlings are crack-prone, and often need repair. Such repairs should be made as soon as a crack is discovered.

Fiberglass Surface Repairs:

1. Remove the fairing, and drill a hole at the end of the crack to stop the crack from progressively getting larger.

2. Clean the inside and outside of the surface to be repaired.

3. Dust the interior surface with baking soda, position a small piece of fiberglass cloth over the crack, and saturate the entire area with cyanoacrylate. (Cyanoacrylate is the basis of the one-drop space-age glues.)

4. Repeat step three.

5. From the outside, fill the crack with baking soda, and harden with cyanoacrylate.

6. Sand smooth, and repaint.

Metal Surface Repairs:

1. Remove the fairing and drill a stop-hole at the end of the crack to stop it from increasing in size.

2. Cut a small piece of metal of the same type as the fairing being repaired. This piece should extend about one inch out from the crack in all directions.

3. Place the patch over the damaged area and drill holes through the patch and fairing. Rivet the patch into place.

4. Touch up with paint as needed.

Refinishing

The FARs allow refinishing decorative coatings of the fuselage, wing, and tail-group surfaces (excluding balanced control surfaces), fairings, cowlings, landing gear, cabin or cockpit interior when removal or disassembly of any primary structure or operating system is not required.

Paint Touch-up: The touching up of small areas on the wings or fuselage of an airplane is very easy, and really makes marked improvements in the plane's general appearance. Touch-up is as follows:

1. Thoroughly wash the area to be touched up. All preservatives such as wax and silicone products must be removed.

2. All loose or flaking paint must be removed. Carefully use very fine sandpaper for this purpose. Do not sand the bare metal.

3. Exposed bare metal must be primed with an aircraft-type zinc chromate primer.

4. Using sweeping spray strokes, apply at least two coats of touch-up paint in a color matching the original surface.

For custom-packaged touch-up paints in spray cans, contact:

> Custom Aerosol Products, Inc.
> P.O. Box 1014
> Allen, TX 75002
> Phone: (214) 727-6912.

Complete Repainting: Although most owners would never attempt to paint their airplane, they should know just what a good paint job consists of. The following information is general in nature and applies to all current repainting methods.

Paint Stripping. Aircraft paint removers are fast-acting water-washable products designed for use on aircraft aluminum surfaces. While using removers, always wear rubber gloves and safety glasses. If remover gets on your skin, flush with plenty of water. If any remover comes in contact with your eyes, flood repeatedly with water and call a physician. Have adequate ventilation.

Do not let remover come in contact with any fiberglass components of the aircraft, such as wingtips, fairings, etc. Make sure that these parts are well masked or removed from the aircraft while stripping is in progress.

Apply the remover liberally, by brush, to the metal surface. When brushing, be sure to brush only in one direction. Keep the surface wet with remover. If an area dries before the paint film softens or wrinkles, apply more remover. It is sometimes advisable to lay an inexpensive polyethylene dropcloth over the applied remover in order to hold the solvents longer. This gives more time for penetration of the film. After the paint softens and wrinkles, use a pressure water hose to thoroughly flush off all residue.

In the case of an acrylic lacquer finish, the remover will not wrinkle the film but only soften it. A rubber squeegee or stiff-bristle brush can be used to help remove this type of finish.

After all paint has been removed, flush the entire aircraft off with a pressure water hose. Let dry.

Using clean cotton rags, wipe all surfaces thoroughly with methyl ethyl ketone (MEK).

Corrosion Removal. After paint stripping, any traces of corrosion on the aluminum surface must be removed with fine sandpaper (no emery cloth), aluminum wool, or a ScotchBrite pad. Never use steel wool or a steel brush, as bits of steel will embed in the aluminum, causing additional corrosion.

Metal Pretreatment. In the case of an aircraft that has been stripped of its previous coating, make sure that all traces of paint and paint remover residue have been removed. Give special attention to areas such as seams and around rivet heads.

The aircraft should be flushed with plenty of clean water to ensure removal of all contaminants. Let dry. Using clean cotton rags, wipe all surfaces thoroughly with MEK.

Apply a metal pretreatment liberally to all the aluminum surfaces of the aircraft. The pretreatment may be applied with clean rags or a brush. While keeping these surfaces thoroughly wet with the pretreatment, scrub briskly with a plastic kitchen scrub pad. You must wear rubber gloves and protect your eyes from splash during this procedure.

After the entire aircraft has been treated, flush very thoroughly with plenty of clean water. Let dry.

The next step is to thoroughly wipe down the entire

aluminum surface with MEK using clean cotton rags. This will ensure that all contaminants are removed prior to application of the primer coat.

Let dry and tack-rag all surfaces.

Priming. The best aircraft primers available today are the two-part epoxy primers. They are specifically designed for aircraft and afford the best in corrosion protection. Epoxy primers give the best adhesion possible both to the substrata aluminum surface and to the finish top coat.

Mix the components of the epoxy primer per the manufacturer's instructions, then let stand for 15 to 20 minutes prior to starting application.

After mixture, pot life is limited.

Care must be taken that only enough primer is used to prime the surface evenly to about 0.0005 in. (½ mil) film thickness. This means that the aluminum substrate should show through with a light yellow coating of the primer coloring the metal.

Drying time of the primer will vary slightly due to the effects of temperature and relative humidity at the time of application. As a general rule, primer should be ready for application of the finish coat within four to six hours.

After the primer is thoroughly dry, wipe the entire surface with clean, soft, cotton rags using a little pressure, as in polishing. Next, tack-rag the entire surface.

You are now ready for application of the top coat finishing system you have selected.

Conditions for Painting. For optimum results, temperature and humidity should be within the following limits:

- ☐ Relative Humidity—20% to 60%

- ☐ Temperature—$\geq$ 70°F.

Departure from these limits could result in various application or finish problems.

Drying time of the finish coating will vary with temperature, humidity, amount of thinner used, and thickness of paint film.

Painting Safety Tips.

- ☐ Ground the surface you are painting or sanding.
- ☐ Do not use an electric drill to mix dope or paint.
- ☐ Wear leather-soled shoes in the painting area.

- [ ] Wear cotton clothes while painting.
- [ ] Keep solvent-soaked rags in a fireproof safety container.
- [ ] Keep spray area and floor clean and free of dust buildup.
- [ ] Have adequate ventilation. Do not allow spray mist or paint fumes to build up in a confined area.
- [ ] Do not smoke or have any type of open flame in the area.

Application of the Top Coat. There are several top coat systems available for use, but not all are suited for owner application or will give adequate long term results.

Polyurethane enamels are the finest aircraft finishes available today. They offer such important characteristics as superior gloss, excellent color retention, and resistance to abrasion, chemical damage, fuel staining, hydraulic fluid spills, and thermal shock.

These characteristics will remain with little or no maintenance over many years of active flying and outside storage. To apply polyurethane enamel, proceed as follows:

Mix the several parts of the polyurethane enamel per the manufacturer's instructions (pot life after mixing is approximately six hours, but will vary with temperature and humidity).

Spray a relatively light tack coat as the first application. Let dry for approximately 30 minutes.

The second coat is applied as a full wet cross-coat. Care should be taken that too much paint is not being applied, resulting in "runs" or "sags."

An overnight dry is preferable before masking for trim color application.

After masking, but before applying the trim color, lightly scuff the trim color surface using the #400 wet-or-dry sandpaper.

Tack-rag, then spray the trim color. Remove the masking as soon as the paint has started to "set."

Acrylic lacquers have been used for many years by some of the largest manufacturers of aircraft and automobiles. These are proven paints, with outstanding durability, good color, and excellent gloss retention characteristics. They are generally easier to apply than polyurethane paints.

Mix the paint and associated other elements per the manufacturer's instructions.

Spray a relatively light tack coat as the first application Let dry for approximately 30 minutes.

Follow this first coat with at least three full wet cross-coats, letting each dry for approximately 30 minutes between coats.

If the material is applied too heavily, orange peel or pinholes are likely to appear.

An overnight dry is preferable before masking for trim color application.

Remove the masking as soon as the trim paint has started to "set."

Non-Structural Fasteners

Seen all those rusted screws on an airplane? It's easy to replace them with non-rusting stainless steel screws. The use of stainless screws on airplanes makes sense, as they don't rust and stain the surrounding area as do the stock screws.

Kits containing the proper size and number of screws recommended for this job (Fig. 12-3) are available from Sporty's, Air Components, Wil Neubert Aircraft Supply, J&M Aircraft, and many FBOs. The kits are also available direct from:

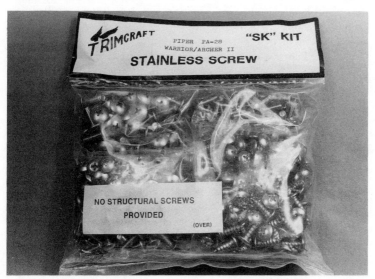

Fig. 12-3. Complete stainless screw kit for a late-model PA-28.

Trimcraft Aviation
P.O. Box 488
Genoa City, WI 53128
Phone: (414) 279-6896

Be *very* careful that you don't strip out the screw holes when removing or installing screws.

Upholstery

You may repair upholstery, decorative furnishings, and seats of the cabin or cockpit provided that such work does not require disassembly of any primary structure or operating system or affect the primary structure of the aircraft.

Due to the wide selection of materials and styles available, I recommend you either contract the job with a professional aviation interior shop, or contact a supplier of complete interiors or slipcovers. The latter method is recommended for the hands-on types who want to save money and do their own work. One such supplier is:

Cooper Aviation Supply Co.
2149 E. Pratt Blvd.
Elk Grove Village, IL 60007

Check *Trade-A-Plane* for listings of other suppliers.

Re-covering the seats is generally easy to accomplish; however, don't attempt to re-cover the seats while they're in the airplane. Remove the seats and do your work in a well-well-ventilated area, or the adhesives you'll be working with will send you flying without the airplane.

Another alternative for seats and carpets is the aircraft wrecking yards. They too advertise in *Trade-A-Plane*.

An entire seat, or structural seat parts, may be replaced, providing approval is met. "Approval," in this case, means using a complete used seat from an identical airplane or factory parts.

Side Windows

Provided the work does not interfere with the structure or any operating system (such as controls and electrical equipment), side windows may be replaced.

Removal: Remove the trim and interior retainer moldings and pull the window from the frame. A replacement window can be made using the original as a pattern, or one may be purchased from a supplier.

Installation: Apply vinyl foam tape to both sides of the window along the outer edges and a sealant around the window at all attachment flanges. Insert the window and install the retainer mouldings. Remove excess tape and sealer from view.

Safety Belts

Safety belts must be replaced when they become frayed, are cut, or the latches become defective. Attaching hardware must be replaced if faulty. Use only approved safety belts.

Other Preventive Maintenance

The following also fall under the heading of preventive maintenance:

- ☐ Troubleshooting and repairing broken landing-light wiring circuits.

- ☐ Replacing bulbs, reflectors, and lenses of position and landing lights.

- ☐ Replacing wheels and skis where no weight-and-balance computation is required.

- ☐ Replacing any cowling not requiring removal of the propeller or disconnection of flight controls.

- ☐ Replacing any hose connection except hydraulic connections.

- ☐ Replacing prefabricated fuel lines.

- ☐ Replacing defective safety wiring or cotter pins.

TOOLS

A few *quality* tools will allow the owner to perform maintenance on his airplane. Cheap tools will frustrate you at the least, and cause damage at worst. Remember the old cliche, ''You get what you pay for.'' A good selection of tools is:

- ☐ Multipurpose knife (Swiss Army knife)
- ☐ ⅜″ ratchet drive with a flex head as an option
- ☐ 2-, 4-, and 6-inch ⅜″ extensions
- ☐ Socket wrench set (⅜″ to ¾″ in ¹⁄₁₆″ increments)
- ☐ 6″ crescent wrench
- ☐ 10″ monkey wrench
- ☐ 6- or 12-point closed (box) wrenches from ⅜″ to ¾″ (In

addition, I recommend a set of open-end wrenches of the same dimensions.)

☐ Pair of channel lock pliers (medium size)
☐ Phillips screwdriver set (three common sizes)
☐ Blade screwdriver set (2″ to 8″ length)
☐ Plastic electrical tape
☐ Container of assorted nuts and bolts
☐ Spare set of spark plugs
☐ A carrying bag or box to keep tools in order and protected (Fig. 12-4)

LOGBOOK REQUIREMENTS

FAR 43.9 requires that entries be made in the appropriate logbook whenever preventive maintenance is performed. An aircraft cannot legally be returned to service without such an entry. A logbook entry must include:

☐ Description of work done
☐ How the work was done
☐ Date work is completed
☐ Kind of airman certificate exercised
☐ Name of the person doing the work
☐ Approval for return to service (signature and certificate number) by the pilot approving the work

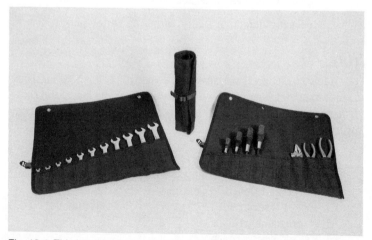

Fig. 12-4. This handy roll-up tool caddy is available from Helsper Sewing Co., 80 Highbury Dr., Elgin, IL 60120. (courtesy Helsper Sewing Co.)

REPLACEMENT PARTS

Many parts needed for proper maintenance and operation of the Piper Indians are very difficult to obtain. If you find yourself in need of a hard-to-find item, call AV-PAC at (800) 228-1836. AV-PAC, a part of Duncan Aviation, Inc., Lincoln, NE, is a warehouse of hard-to-find Piper parts.

For an in-depth study of preventive maintenance, check your local library for a copy of the now out-of-print *Lightplane Owner's Maintenance Guide,* by Cliff Dossey.

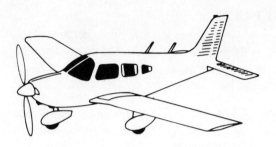

Chapter 13

Engine Operations

Owners of the older Piper Indians have recently been voicing their concerns about the limited availability of 80/87 grade fuel and the use of the higher-leaded 100LL in its stead.

AVGAS PROBLEMS

Around the country, supplies of 80/87 grade aviation fuel have dwindled to almost nothing. This is because most refiners are not making 80/87 grade aviation fuel. The overall trend in aviation fuel availability is towards a complete phaseout of 80/87 aviation grade fuel.

The refiners claim there is too little profit in 80/87 production. They further state that the new 100LL avgas is an adequate replacement for the 80/87 grade. What the fuel refiners *really* mean is that they are meeting the needs of supply and demand . . . the customer will use the supply that the supplier demands him to use!

Color Coding of Avgas

Avgas is color-coded to preclude introducing the incorrect fuel into an aircraft when refueling.

☐ Red: 80 octane containing .50ml lead/gal.

☐ Blue: 100 octane containing 2ml lead/gal.

☐ Green: 100 octane containing 3ml lead/gal.

It's interesting to note that "Blue 100," also referred to as 100 LL (LL for low lead), contains four times the amount of lead as 80/87. Gradually, 100LL will become the only fuel available for small piston airplane engines. It must be used as a replacement fuel whenever 80/87 is not available.

Low Lead is Blue

The color of 100LL is blue; perhaps this is appropriate. The continuous use of 100LL as a replacement for 80/87 causes fouling deposits in the combustion chamber and the oil. This results in increased spark plug maintenance and frequent oil changes, as well as other repairs.

Lead-related fouling can be controlled to some extent by the method of engine operation. The following information is directed towards the operators of non-fuel-injected Lycoming engines; however, it should also be of value to operators of other engines.

To keep engine deposits at a minimum when using the 100 LL, pilots need to understand the proper techniques for fuel mix leaning and engine shutdown.

LEANING THE FUEL MIX

Engine operation at full rich mixture causes carbon fouling and allows excessive lead buildup in the combustion chamber. It is very important to learn, and use, proper fuel mix leaning techniques.

Spark Plug Heat Range

The environment in which a spark plug operates changes as engine temperature and fuel-air ratios (fuel mix) change. At low rpm an engine runs very rich. Low-power operations, such as taxiing and other ground operations, cause the ceramic tip of the spark plug to run "cold." When the tip is cold, it can accumulate carbon deposits that can lead to engine failure during the next application of full power.

Medium temperature ranges are those normally associated with cruise power settings. During cruise it is possible to accumulate some lead buildup. However, this can be controlled by proper leaning techniques.

Takeoff power represents the highest temperatures at which the spark plug must operate. No deposits form during high-temperature operation, in fact, some deposits may burn off.

A spark plug is sensitive to fuel-air ratios because changes in the ratio greatly change combustion chamber operating temperatures.

Spark plugs are available in various heat ranges from "hot" to "cold." The correct spark plug for an engine is one that is hot enough to prevent hard carbon buildup to form, yet cold enough to avoid pre-ignition.

The heat range of a spark plug is controlled by the rate of heat transfer from the spark plug to the engine head. A hot plug transfers combustion chamber heat slowly to the cylinder head. A cold plug makes this transfer quickly.

General Leaning Rules

The object of leaning is to reduce engine deposits and increase fuel efficiency. Unless your owner's manual indicates otherwise, follow these general rules when leaning your fuel mixture:

1. Never lean the mixture from full rich during takeoff, climb, or high-performance cruise operation. (*Exception*: During takeoff from high-elevation airports or during climb at higher altitudes, leaning may be required to eliminate the roughness or the reduction of power that might occur at full rich mixtures. In such instances, the mixture should be adjusted only enough to obtain smooth engine operation.)

2. Careful observation of temperature instruments should be practiced.

3. Always return the mixture to full rich before increasing power settings.

4. During the approach and landing sequence, the mixture should be placed in the full rich position (unless landing at high-elevation fields, where leaning may be necessary).

Leaning Procedures

Each basic type of fuel metering system will require a slightly different method of leaning.

Float Carburetor. At cruising altitude and speed, adjust the mixture control to maximum engine rpm (just prior to engine roughness). The roughness observed as the point of maximum rpm is passed is caused by detonation.

At cruise, lean the mixture until roughness occurs, then enrich slightly until the engine is again operating smoothly. A slight increase in airspeed may be noted at the proper leaning point.

Fuel Injection. At cruise, the initial leaning is done by adjusting the mixture control for maximum rpm (just prior to engine roughness). Final setting is done with reference to an exhaust gas temperature (EGT) gauge. As a rule of thumb, always set your mixture 50 degrees on the rich side of maximum EGT. Consult your aircraft operations manual for the maximum EGT.

Turbochargers. Before leaning any turbocharged engine, consult the airplane's operations manual. The manual will give maximum temperatures for the turbine inlet and cylinder head. Actual leaning must be accomplished with reference to the TIT (turbine inlet temperature) and CHT (cylinder head temperature) gauges. As a rule of thumb, never exceed a CHT of 435°.

Engine Monitoring

All pilots are familiar with monitoring engine rpm, oil temperature, and oil pressure. However, as airplanes become complex, so do the systems requiring monitoring.

Exhaust Gas Temperature (EGT). Exhaust gas temperatures vary with the fuel mixture:

Rich mixture = low EGT

Lean mixture = high EGT

An EGT gauge samples the temperature of the exhaust gases as they enter the exhaust manifold (Fig. 13-1). The EGT gauge is extremely valuable for monitoring leaning procedures; however, it is limited in accuracy with float-carburetor engines due to the uneven distribution of the fuel mixture through the intake manifold. In fuel-injected engines, this distribution is extremely precise; hence, the EGT gauge will be accurate.

Cylinder Head Temperature (CHT). Cylinder head temperatures are monitored with a CHT gauge that remotely indicates the temperature of the cylinder heads (Fig. 13-2). Problems such as inadequate engine cooling can be detected by its use.

Carburetor Temperature. Proper monitoring of carburetor throat temperature is necessary to prevent ice from forming in the carburetor (Fig. 13-3).

Fig. 13-1. This digital readout unit displays EGT, oil, and CHT. (courtesy Electronics International)

Carburetor Ice Detector. The ice detector is designed to actually detect ice, not just low carburetor throat temperature. Ice is a product of temperature and humidity. The mere indication of carburetor temperature is not satisfactory, as it does not relate the

Fig. 13-2. The graphic display shown on this unit allows instant comparison of cylinders. (courtesy Insight Instrument Corp.)

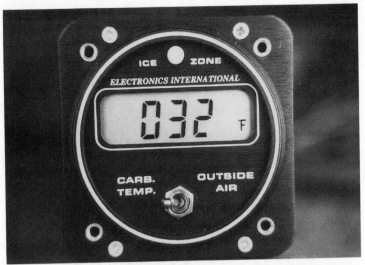

Fig. 13-3. Digital display of carburetor throat temperature and outside air temperature (OAT). (courtesy Electronics International)

complete picture (Fig. 13-4). The detector utilizes an optical probe in the carburetor throat and is so sensitive that it can detect ''frost'' up to five minutes before ice begins to form, allowing the pilot plenty of time to take corrective action.

ENGINE SHUTDOWN TECHNIQUE

The engine deposit formation rate can be greatly reduced by controlling ground operations to minimize the separation of non-volatile components of the higher-leaded aviation fuels. The formation rate is accelerated by low fuel-mixture temperatures caused by the rich fuel-air mixtures associated with idling and taxiing operations. This was also noted in the reference to spark plug operating temperatures.

To reduce the effects of this high deposit formation environment, it is essential that engine idling speeds be set in the 600 to 650 rpm range with the idle mixture adjusted properly to provide smooth idling operation.

Engine speed should be increased to 1200 rpm for one minute prior to shutdown. This will increase the combustion chamber temperature and allow the deposits to dissipate. After one minute, slow the engine down and lean until operation ceases.

Fig. 13-4. The function of the ice detector is to warn the pilot of actual ice formation inside the carburetor throat. This particular unit, available as an STC, was also available as an option on many models of Piper aircraft. (courtesy ARP)

CHEMICAL HELP TO FIGHT LEAD

In April 1977, the use of Tricresyl Phosphate (TCP) was approved by the FAA for use in Lycoming and Continental engines that do not incorporate turbosuperchargers.

TCP is a fuel additive that is used to prevent lead fouling. It is available from most FBOs and:

> Alcor, Inc.
> 10130 Jones-Maltsberger Rd.
> Box 32516
> San Antonio, TX 73284

FROM GROUND TO AIR

As a result of the scarcity of 80/87 avgas, there has been considerable controversy and discussion about the use of ''auto'' fuels (sometimes referred to as *mogas*) in lieu of the 100LL products in aircraft engines.

The use of auto fuels has many pros and cons. The debate has been going on for many years, and will probably continue until piston airplane engines are no longer used.

It is up to the individual pilot to make his own choice about the use of non-aviation fuels in his airplane. To aid in making a decision consider:

PRO:

☐ Unleaded auto fuel is less expensive than 100LL avgas.

☐ Based on extensive testing, it appears that auto fuels operate well in the older engines that require 80/87 octane fuel.

☐ If you own a private gas tank/pump, it will be advantageous to utilize auto fuel. It'll be far easier to locate an auto fuel supplier willing to keep your tank filled than it will be to find an avgas supplier willing to make small deliveries. This could be the deciding factor at a small private strip.

CON:

☐ There is a decided lack of consistency among the various brands of auto fuels (gasolines) and their additives. In particular, many low-lead auto fuels have alcohol in them. Alcohol is destructive to some parts of the typical aircraft fuel system. A similar problem has been noted with older automobiles.

☐ The engine manufacturers (Lycoming and Continental) claim the use of auto fuels will void warranty service. This is not really important unless you have a new or factory-remanufactured engine.

☐ Many FBOs are reluctant to make auto fuels available for reasons such as product liability and lower profit. However, a look in *Sport Aviation*, the Experimental Aircraft Association (EAA) magazine, indicates an upward trend in the number of FBOs selling mogas.

Mogas STCs

Mogas STCs are available from the EAA for many Piper Indians at a minimal fee. The STCs allow the legal use of unleaded regular automobile gasoline manufactured to the ASTM Specification D-439 (American Society for Testing Materials, 1916 Race St., Philadelphia, PA 19103).

Not all states comply with the ASTM specifications. The following is a list of states requiring compliance with this standard for automobile gasolines:

Arizona	Maryland
Arkansas	Montana
Alabama	Nebraska
California	Nevada
Colorado	New Mexico
Connecticut	New York
Florida	North Carolina
Georgia	South Carolina
Hawaii	North Dakota
Idaho	South Dakota
Iowa	Oklahoma
Indiana	Rhode Island
Kansas	Texas
Louisiana	Tennessee
Maine	Utah
Massachusetts	Virginia
Minnesota	Wisconsin
Mississippi	Wyoming

Mogas Warning

A note of advice: Prior to purchasing the auto fuel STC, check with your insurance carrier and get their approval . . . *in writing*.

Some aviation insurance companies are not keen on the use of mogas, and if you don't comply with their wishes, they could cancel your insurance, raise your already-too-high rates, or even deny a claim.

If you desire further information about the legal use of auto fuels in your airplane, contact the EAA. This organization has an ongoing program of testing airplanes and obtaining STCs for the use of auto fuel. Contact the EAA at:

EAA-STC
Wittman Airfield
Oshkosh, WI 54903

Fuel Transportation

If an FBO does not have mogas , the airplane owner must find a method of safely transporting the fuel to his airplane.

The most dangerous method of transporting fuel is to carry five-gallon cans in the trunk of a car. This really makes you a "ground-running Kamikaze" looking for a place to incinerate yourself (and others).

The best method—also the most expensive—is to have auxiliary fuel tank(s) installed on your van or truck.

Once you get the fuel to the airport, you may find you have a problem with the FBO. He may say that you cannot fuel your own airplane, citing reasons such as safety, local ordinance, his insurance, etc. Some FBOs have even gone to the trouble of making aircraft parking areas inaccessible to ground vehicles (except their fuel tankers).

To fuel the fire even further: Re-read the partial reprint of FAA Advisory Circular 150/5190 in Chapter 12 of this book. Be familiar with the contents, as you may someday need to quote from it to stand up for your rights!

Chapter 14

Propellers

Most pilots don't give much thought to the metal propeller up there on the front of the engine. They should! Even though a high margin of safety is built into the design of metal propeller blades, failures do occur. Reports of failures cannot be attributed to any particular engine/propeller/aircraft combination.

WHY PROPELLERS FAIL

Propeller blade failures occur because of fatigue cracks that start as dents, cuts, scars, scratches, nicks, or leading edge pits. Only in rare instances has failure been caused by material defects of surface discontinuities existing before the blades were placed in service.

Fatigue Failure

Fatigue failures of blades have occurred at the place where previous damage has been repaired. This may be due to the repair merely amplifying the problem. Too much flexing of the blade, such as in blade straightening or blade pitching operations, can overstress the metal, causing it to fail.

Propeller Flutter

Flutter, a vibration causing the ends of the blade to twist back and forth at a high frequency around an axis perpendicular to the crankshaft, can cause blade failure.

At certain engine speeds, this vibration can become critical, and if the propeller is allowed to continue to operate in this range, blade failure can occur. At the very least, metal fatigue will result. It is for this reason that tachometer accuracy is so very important.

Periodic tachometer accuracy checks should be made using reliable testing instruments. Then, by referencing your tachometer, you can avoid propeller speeds that can be damaging. These speeds are indicated as red arcs on the tachometer, and are listed in the operations manual.

Hidden Problems

Metal propeller blade failure can also occur in areas seldom inspected, such as under leading edge abrasion boots and under propeller blade decals. Inspect these hidden areas when the propeller is serviced/repaired.

HOW PROPELLERS FAIL

There are many stresses on a propeller. The propeller is at the end of the energy chain, and is responsible for efficiently converting engine power into thrust. During normal operation, four separate stresses are imposed on a propeller:

☐ Thrust

☐ Torque

☐ Centrifugal force

☐ Aerodynamic force

The stresses that occur in the propeller blades may be viewed as parallel lines of force that run within the blade approximately parallel to the surface. Additional stresses are imposed by flutter or uneven tracking of the blades.

When a defect occurs (scratch, nick, dent), these lines of force are squeezed together, concentrating the stress. The increase in stress can be sufficient to cause a crack to start, which results in an even greater stress concentration. This greater stress concentration causes the crack to enlarge until blade failure occurs (Fig. 14-1).

Most blade failures occur within a few inches of the blade tip. However, failures can occur in other portions of the blade when

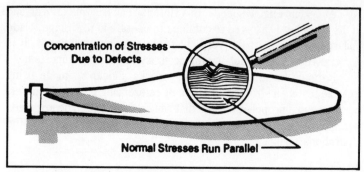

Fig. 14-1. An example of how small defects cause concentration of stress leading to blade failure. (courtesy FAA)

dents, cuts, scratches, or nicks are ignored. *No* damage should be overlooked or allowed to go without repair.

PROPELLER CARE

Propellers can be quite complicated. The typical two-bladed constant-speed propeller is made up of approximately 200 separate parts. Naturally, a fixed-pitch prop has only one part.

All propellers will wear and become unairworthy at some point in time. Propellers are expensive to have repaired—$2,000 for a typical constant-speed prop. The best way to prolong propeller life is through regular care and maintenance.

Blade Care

☐ Keep the blades clean, as complete inspections cannot be made if blades are covered with dirt, oil, or other foreign matter.

☐ Avoid engine run-up areas containing loose sand, stones, gravel, or broken asphalt. These particles can nick the propeller, if sucked into it during run-up.

☐ *Propeller blades are not designed to be used as handles for moving an airplane.*

☐ During the normal 100-hour or annual inspection, the engine tachometer should be checked for accuracy to preclude operation in any restricted rpm range.

☐ During inspections, check applicable ADs, and comply with them.

Propeller Inspection

When performing a preflight inspection, check not only the leading edge of the propeller, but the entire blade for erosion, scratches, nicks, and cracks. Regardless of how small any surface irregularities appear, consider each as a stress point subject to fatigue failure.

PROPELLER BALANCING

An out-of-balance propeller causes vibration, which results in the premature failure of aircraft components. Vibration results in the "shaking apart" of anything mechanical. If you have ever driven a car with a tire out of balance, and felt the throbbing vibration caused by that condition, you have an idea what an out-of-balance propeller can do.

Dynamic balancing of a propeller is quite similar to high-speed on-the-car balancing of a car tire. The measuring is done with a strobe light, tachometer, and vibration detector. Corrective weights are placed on the propeller hub.

Propeller balancing has long been done on turboprop engines, but it is not routinely done on piston engines.

For further information about propeller balancing, contact your local FBO or:

Chadwick-Helmuth Company, Inc.
4601 N. Arden Dr.
El Monte, CA 91731
Phone: (818)575-6161

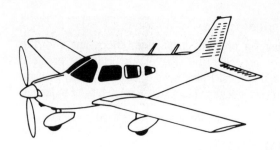

Chapter 15

Avionics

The Piper Indian airplanes are very versatile transporters of people. However, to get all from them that they can provide, you may have to upgrade the avionics—particularly on the older birds, or those minimally equipped.

TYPES OF AVIONICS

A-panel: Audio panel. Allows centralized control of all radio equipment (Fig. 15-1).

ADF: Automatic Direction Finder (Fig. 15-2).

CDI: Course Deviation Indicator is panel-mounted and gives a visual output of the NAV radio (Fig. 15-3).

COMM: VHF transceiver for voice radio communications (Figs. 15-4, 15-5).

DME: Distance Measuring Equipment (Fig. 15-6).

ELT: Emergency Locator Transmitter (required by FARs for all but local flying).

LOC/GS: Localizer/Glide Slope. Visual output is via CDI, with the addition of a horizontal indicator to show glide path (Fig. 15-7).

LORAN C: A system of very accurate radio/computer-based navigation completely separate from the VOR-based systems (Figs. 15-8, 15-9).

MBR: Marker Beacon Receiver (Fig. 15-10).

NAV: VHF navigation receiver for utilizing VORs.

Fig. 15-1. The purpose of an audio panel is to place speaker, headphone, and microphone controls in one central location. (courtesy Collins General Aviation Division of Rockwell International)

Fig. 15-2. ADF with digital display of frequency and time. (courtesy NARCO Avionics)

NAV/COMM: Combination of COMM and NAV in one unit (Figs. 15-11 through 15-12).

RNAV: Random Area Navigation, a microprocessor-based system that allows considerable flexibility in course planning and flying (Fig. 15-13).

XPNDR: Transponder; may or may not have altitude encoding capability (Fig. 15-14).

FLYING NEEDS

If you are a casual flier, and seldom make cross-country flights, you can get by with a minimum of equipment:

☐ NAV/COMM
☐ XPNDR
☐ ELT

Fig. 15-3. King KI 205, CDI for VOR NAV receiver. (courtesy Bendix/King Radio)

Fig. 15-4. Apollo 706 COMM radio is full-featured, with digital entry and frequency memories. (courtesy II Morrow)

Fig. 15-5. Collins Micro Line II COMM radio. Notice the active/standby switch on the bottom right. (courtesy Collins General Aviation Division of Rockwell International)

Fig. 15-6. Collins DME-451. This digital readout indicates the air-to-ground distance from a VOR in miles and time, and your current speed. (courtesy Collins General Aviation Division of Rockwell International)

Fig. 15-7. Collins IND-351. Notice the horizontal needle for glide slope indication. (courtesy Collins General Aviation Division of Rockwell International)

Fig. 15-8. This NARCO LORAN has a membrane-touch control panel with a digital display of all navigational information. (courtesy NARCO Avionics)

It wasn't too many years ago that most cross-country flying was done by pilotage (reading charts and looking out the windows for checkpoints). But today's aviator has become accustomed to the advantages of modern navigation systems. Therefore, I feel that an airplane equipped in this minimal fashion would be inappropriate for the typical pilot of today.

If you do a lot of VFR cross-country flying, as most family pilots do, you will need a little more equipment to ease your workload, and to give you backup in case of failure:

☐ Dual NAV/COMM   ☐ XPNDR

☐ DME   ☐ ELT

☐ RNAV   ☐ LORAN C (can effectively

☐ ADF   replace DME, ADF, and RNAV)

Fig. 15-9. Apollo I LORAN. (courtesy II Morrow)

Fig. 15-10. This segmented Marker Beacon Receiver indicator will light as you pass over the markers on an ILS approach. (courtesy Bendix/King Radio)

Fig. 15-11. The Mark 12 D. Notice the small buttons just below each digital display; These allow you to exchange the active and standby frequencies. (courtesy NARCO Avionics)

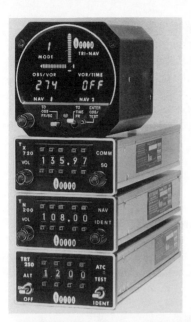

Fig. 15-12. The Terra stackables. Notice the CDI, it is all solid state with digital display. (courtesy Terra)

Fig. 15-13. Collins RNAV. The readout reads: waypoint #5, on the 340-degree radial, 120 miles out. (courtesy Collins General Aviation Division of Rockwell International)

Fig. 15-14. The King transponder, KT-79—all digital, with digital altitude indicator. (courtesy Bendix/King Radio)

243

The additional NAV/COMM can be used in easing your workload, as well as providing backup in case of partial equipment failure. If you want to add one more thing that will really ease your workload on long cross-country trips, get an autopilot. It doesn't have to be a complex model; even a wing leveler would be a great assist.

IFR flying (and the Piper Indians are certainly capable of it) will require still more equipment. In addition to the above, you must add:

☐ LOC/GS ☐ XPNDR (altitude reporting)
☐ MBR ☐ Autopilot (optional)

The entire IFR installation must be certified, so be prepared to spend lots of money if you are planning to completely re-outfit a plane for IFR. It might be worthwhile to entertain thoughts of changing airplanes, rather than just avionics. Often you can purchase a newer airplane, equipped as you want, cheaper than you can update your present airplane.

UPDATING YOUR AIRPLANE

There are several ways of filling those vacant spots and replacing older equipment on your instrument panel. Some are more expensive than others.

New Equipment

New equipment is state-of-the-art, offering the newest innovations, best reliability, and—most important—a warranty. An additional benefit is the fact that the new solid-state electronic units are physically smaller and draw considerably less electric power than older equipment. This is extremely important for the person wanting a "full panel." New avionics can be purchased from your local avionics dealer or a discount house.

You can purchase all the equipment from your local dealer and have him install it. This will be the most expensive route you can take when upgrading your avionics; however, in the long run it could be the most cost-effective. You'll have new equipment, expert installation, and service backup. You will also have someone nearby you can "discuss" problems with.

The discount house will be considerably cheaper for the initial purchase, but you may be left out in the cold when the need for warranty service arises. Some manufacturers will not honor warranty service requests unless the equipment was purchased from, and installed by, an authorized dealer. Discount houses advertise heavily in *Trade-A-Plane*.

Used Equipment

Used avionics can be purchased from dealers or individuals. The aviation magazines and *Trade-A-Plane* are good sources of used equipment—however, a few words of caution about used avionics. Purchase *nothing:*

- ☐ with tubes in it.
- ☐ more than six years old.
- ☐ made by a defunct manufacturer (parts could be difficult to obtain).
- ☐ "as is."
- ☐ "working when removed."

Good used avionics equipment can be a wise investment, but judging the good from the bad is very risky, unless you happen to be an avionics technician or have access to one.

I recommend against the purchase of used avionics, unless you are *very* familiar with the source. Even then, I would not recommend used primary IFR equipment.

Reconditioned Equipment

There are several companies that advertise reconditioned avionics at bargain—or at least low—prices.

The equipment has been completely checked out by an avionics shop. Parts that have failed, are near failure, or are likely to fail, will have been replaced.

These radios offer a fair buy for the airplane owner and are usually warranted by the seller, but be advised that a reconditioned radio is not new. Everything in the unit has been used, but not everything will have been replaced during reconditioning. You will have some new parts and some old parts. As long as you are aware of this drawback, I feel reconditioned equipment makes sense if you are budget-minded, provided you have no desire for the latest "bells and whistles." Also, few pieces of reconditioned equipment will exceed six or seven years of age.

HANDY BACKUP

Want an extra COMM radio for backup, and don't want to spend a thousand dollars to get it? Sounds like most of us, always wanting something for nothing.

Well, they're not free, but almost (in airplane dollars, that is). The simplest backup COMM radio is a handheld (Figs. 15-15, 15-16.)

The typical handheld portable COMM radio of today has 720-channel capability with pushbutton frequency entry, memory functions, rechargeable nicad batteries, etc.

The HT, as it is often called, requires no installation, mechanical or electrical. It is independent of the airplane's electrical system, unless you want to plug it into the cigarette lighter.

Pricewise, I don't think an HT can be beaten. Most are available for under $500.

RECOMMENDATIONS

My strongest recommendation when contemplating the purchase of additional avionics is to save your money until you can purchase new equipment. New avionics offer more features each model year,

Fig. 15-15. The NARCO HT 830. Notice the digital readout; it says: "From" the VOR on the 063-degree radial. (courtesy NARCO Avionics)

Fig. 15-16. ICOM IC-A2 HT. (courtesy ICOM)

the size goes down, the electrical appetite is reduced, and the reliability factor goes up. Advanced technology even makes them more of a bargain than those of 20 years ago. Complete NAV/COMMs are available for less than $1,000.

Don't even *think* about trading in equipment that is currently working properly. You can't replace it for what a dealer will give you. Keep it as your second system, or sell it outright.

LORAN-C

Long Range Navigation, called LORAN (LORAN-C is the latest version), is based upon low-frequency radio signals, rather than the VHF FAA navaids normally associated with flying. Actually, LORAN was not really intended for general aviation usage, but it has become very popular.

LORAN Advantages

Due to the propagation properties of radio waves at the frequencies it utilizes, LORAN offers distinct advantages over normal VHF navaids (such as VORs).

With LORAN there is no VHF line-of-sight "usable range" limit. This means that unlike most VORs, usable only within 100 miles of the station, LORAN is usable many hundreds of miles from the actual transmitter. This opens up some very interesting possibilities for use.

Much general-aviation activity is conducted at low altitude—under 2,000 feet—and in very isolated areas . . . the boondocks! This low-altitude flying is a very limiting factor when navigating by standard VORs. The low altitude means that the VOR may be of little or no use, as the signals are basically line-of-sight, and you may be blocked from "sight" by mountains or even the horizon. This is where LORAN shines; it is normally usable right down to the ground.

Without going into extensive theory about operation, the LORAN-C unit can, by receiving several LORAN signals at one time and comparing them, determine its exact location *within a few feet*! This will be displayed on the readout as latitude and longitude. Navigation is accomplished by the use of *waypoints*.

Waypoints are geographical locations entered into the LORAN by the operator via the keyboard. The unit will then compare the signals to the geographical inputs and give continuous information concerning course direction, time elapsed, estimated time enroute, distance traveled, distance to destination, etc., based on reference to the waypoints. All this in one box!

The first LORAN-C units seen on the aviation market were reworked marine versions. More recently, several companies have introduced models designed exclusively for aviation.

There are many makes/models of LORAN-C on the market. They vary primarily in the number and type of features found on the individual unit. Many of the first LORAN-Cs installed in small airplanes were not certified for IFR work; however, this does not mean they were incapable or inaccurate. It only means that the manufacturers were unwilling to spend the many dollars necessary for certification. This scene has now changed.

Most LORAN units are now TSOed (Technical Standard Ordered). Price—at one time an indicator of TSO or non-TSO—is dropping quickly, and some TSOed LORANs are available for less than $1,000. Of course, you can spend more money if you wish.

It is also possible to slave the LORAN to the autopilot.

LORAN Limitations

LORAN coverage areas vary in the interior of the United States from the coastal areas. Areas that are adequately covered include:

- ☐ East Coast (from the Mississippi River east)
- ☐ Central U.S. (between the Mississippi River and a line extending from Houston, Texas north to the Canadian border)
- ☐ California
- ☐ Oregon
- ☐ Washington
- ☐ Idaho
- ☐ Nevada
- ☐ Arizona
- ☐ Western parts of the Rocky Mountain states

Coverage information is based on information provided by the U.S. Coast Guard, and efforts are underway to fill in the existing mid-continent coverage gap. By the way, LORAN is operated by the Coast Guard, not the FAA.

SAVE YOUR EARS

After cruising for many hours in a plane, whether it's a Piper or any other make, your ears will be ringing from all the constant noise. It is possible to damage your hearing with constant assaults of loud noise, and the small airplane's cabin is just the right place for such injury to take place.

The FAA has issued Advisory Circular 91-35, partially reprinted here for your information:

****

AC: 91-35

Subject: NOISE, HEARING DAMAGE, AND FATIGUE IN GENERAL AVIATION PILOTS

1. **PURPOSE.** This circular will acquaint pilots with the hazards of regular exposure to cockpit noise. Especially pertinent are piston-engine, fixed-wing, and rotary-wing aircraft.

2. **BACKGROUND.**

    a. Modern general aviation aircraft provide comfort, convenience, and excellent performance. At the same time that the manufacturers have developed more powerful engines, they have given the occupants better noise protection and control, so that today's aircraft are more powerful, yet quieter than ever. Still, the levels of sound associated with powered flight are high enough for general aviation pilots to be concerned about participating in continuous operations without some sort of personal hearing protection.

    b. Most long-time pilots have a mild loss of hearing. Many pilots report unusual amounts of fatigue after flights in particularly noisy aircraft. Many pilots have temporary losses of hearing sensitivity after flights, and many pilots have difficulty understanding transmissions from the ground, especially during critical periods under full power, such as takeoff.

3. **DISCUSSION.**

    Like carbon monoxide, noise exposure has harmful effects that are cumulative—they add together to produce a greater effect on the listener both as sound intensity is increased, and as the length of time he listens is increased. A noise that could cause a mild hearing loss to a man who heard it once a week for a few minutes might make him quite deaf if he worked in it for eight hours.

****

The Fix

As with everything, there is a fix. For the airplane driver and his passengers, there is the headphone intercom system.

Intercom systems come in all types and with varied capabilities.

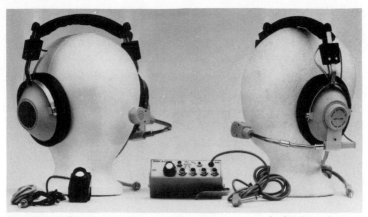

Fig. 15-17. This Hush-A-Com intercom system is completely stand-alone, requiring no installation. (courtesy Revere Electronics of Cleveland, Ohio)

Some are an extension of the audio panel, primarily for the use of the pilot in his duties; others are stand-alone. "Stand-alone" means they are not hooked to anything in the airplane, but rather are completely portable (Figs. 15-17, 15-18).

No matter what type you select, the ear protection will be controlled by the quality of the headphones used. For proper ear protection you must use *full ear cover* headsets, not the lightweight stereo types so popular with the high-school set (Figs. 15-19 through 15-21).

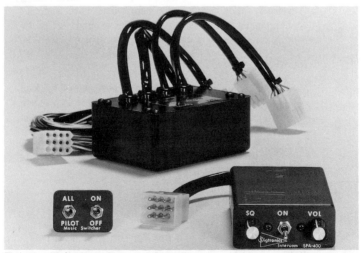

Fig. 15-18. A complete Sigtronics SPA-400 intercom system, capable of carrying radio, voice, and music. (courtesy Sigtronics Corp. of Covina, California)

250

Fig. 15-19. This type of headset is very comfortable to wear over long periods of time. The large muffs will protect your hearing from damage. (courtesy TELEX Communications)

Fig. 15-20. This particular headset not only helps in noise reduction, it has an amplified microphone. (courtesy David Clark Corp.)

There are several manufacturers of adequate headsets, and you will see their ads in the magazines and *Trade-A-Plane*. Don't make a selection based solely on an advertisement. Talk to other pilots, then go to an aviation supply store and try a few. Pay particular attention to the weight, as weight will become a fatigue factor during long flights. After you find the model you like, purchase and *use* it.

Fig. 15-21. This ultralight headset is ideal for an already-quiet cockpit—and it's very comfortable. (courtesy TELEX Communications)

Fig. 15-22. These headsets are ideal for long uncontrolled VFR trips, with musical interludes. . . in stereo. (courtesy Sigtronics)

Fig. 15-23. The VA-1 is an electrical system analyzer, capable of monitoring the aircraft's electrical system. (courtesy Electronics International of Hillsboro, Oregon)

Fig. 15-24. Accurate outside air temperature is shown on this digital display. (courtesy DAVTRON)

Fig. 15-25. An accurate timepiece is necessary when flying. This is the most modern available. (courtesy DAVTRON)

There is another advantage to headsets: Plug them into a stereo and listen to music as you make your flights—only the uncontrolled flights, of course! This sure beats listening to the ADF—or an old portable radio, like I used to do (Fig. 15-22).

OTHER DEVICES

In this the day of modern technology, gadgets, pushbuttons, and digital readouts, some owners have seen fit to add numerous options to the panels of their airplanes.

Among the more common additions are the digital voltmeter, digital outside air temperature (OAT) gauge, and the digital chronometer (Figs. 15-23 through 15-25).

To some, the instrument panel of an airplane is a functional device. To others, it's a statement made by the owner.

In either case, much care must be taken when filling up the panel—not just to provide useful instrumentation, but to plan it functionally, and accomplish it economically.

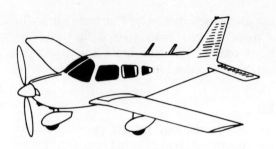

Chapter 16
Hangar Flying

The following pages are devoted to comments, statements, rumors, opinions, and *facts* that are often heard about the Piper airplanes covered in this book.

CLUBS

For the most information about your airplane, join a club that supports your particular model. In the field of Piper Indians, there are two such clubs.

The Cherokee Pilots Association

The Cherokee Pilots Association is devoted to the owners and pilots of:

☐ Cherokee   ☐ Dakota
☐ Warrior   ☐ Cherokee Six
☐ Archer   ☐ Lance
☐ Arrow   ☐ Saratoga

The association publishes a fine monthly newsletter with well-documented information pertinent to the safe/cost-effective ownership of the Piper single-engine airplanes, including:

- [ ] Maintenance and Modification Ideas
- [ ] Exchange of Technical Information
- [ ] Piper Service Letters
- [ ] FAA Airworthiness Directives
- [ ] FAA Airworthiness Alerts

Additionally, the Association provides a group insurance program for your aircraft, a family-oriented annual convention, regional fly-ins, and help as close as a telephone. For more information, contact:

Cherokee Pilots Association
P.O. Box 20187
St. Petersburg, FL 33742
Phone: (813) 576-4922

International Comanche Society

The International Comanche Society, as the name implies, supports owners and operators of the PA-24 Comanche series of airplanes. It has worldwide membership, and many chapters, called "tribes."

A newsletter, the *Comanche Flyer*, is provided to the members on a monthly basis (except August). It includes maintenance and safety information, news of social gatherings, and loads of Comanche-oriented advertising. A membership directory is printed yearly.

For more information about ICS, contact:

International Comanche Society
P.O. Box 477
Frostproof, FL 33843
Phone: (813) 635-5555

Aircraft Owners and Pilots Association

The Aircraft Owners and Pilots Association (AOPA) is an organization that supports all of general aviation.

Membership is open to all (pilots, non-pilots, owners, non-owners, etc.). The AOPA issues a top-rate monthly magazine, the

255

AOPA Pilot, and offers many member services including research, purchasing guidance, title searches, etc. For further information, contact:

Aircraft Owners and Pilots Association
421 Aviation Way
Frederick, MD 21701
Phone: (301) 695-2000

WHO SAYS WHAT

Due to the large number of Piper all-metal airplanes around, just about everyone has something to say about them. Here are some comments I have heard:

Insurance Carriers Say:

"Insuring the PA-28 airplane is easy, as there are no real secrets to them. They are reliable, easy-to-fly, and replacement parts are available everywhere."

"We are glad to insure low-time retractable pilots in the Arrow, because it has the automatic landing gear system. We do not, however, feel the same about Comanche pilots."

"Most of our insureds are weekend fliers and have families flying with them. That's good; it keeps the pilots in line."

"The Warrior wing has reduced stall/spin accidents on PA-28s to almost zero."

Line Boys Say:

"Some of the low-wing Pipers I see come in here are more dolled up than the expensive Beech jobs. You know, curtains at the windows and nice upholstery. Some people really care about their airplanes."

"I like low-wings to refuel; I can't fall off a ladder with them."

"The old PA-28 drivers were sure happy to see red [80/87 octane] fuel again."

"A lot of planes are being fueled from gas cans. The owners have mogas STCs. We may start to sell mogas here soon due to the demand."

"We started selling mogas about six months back; now older Cessnas and Pipers come as far as 50 miles to refuel."

256

Mechanics Say:

"Not much can go wrong with a Piper low-wing that creates any real problems; it's just that there are so many ADs to contend with. But most of them aren't serious."

"The four-cylinder Lycoming 150, 160, and 180-hp engines are very tough, yet not expensive to maintain."

"Watch the front strut for underinflation. If it gets low, you have very little propeller clearance."

"Be sure the larger [½ inch] exhaust valves have been installed in the older engines. This will increase TBO from 1,200 to 2,000 hours."

"These planes are simple enough for the typical owner to care for with little supervision. Owner care is something I like to see. It makes my job easier, 'cause they keep clean airplanes, usually free of minor problems."

NTSB Says:

The following tables of comparative accident data are a compilation from a study made by the National Transportation Safety Board (NTSB). All rates are based on 100,000 hours of flying time.

It's interesting to note where the various makes/models place in these tables. (*Worst* is at the top; *best* is at the bottom.)

If you are unfamiliar with some of the mentioned makes/models, I suggest you consult my book *The Illustrated Buyer's Guide to Used Airplanes* (TAB #2372).

Fatal Accident Rate
Comparison by Manufacturer

Make	Mean Fatal Accident Rate	Make	Mean Fatal Accident Rate
Bellanca	4.84	Mooney	2.50
Grumman	4.13	Piper	2.48
Beech	2.54	Cessna	1.65

Engine Failure

Aircraft	Rate	Aircraft	Rate
Globe GC-1	12.36	Grumman AA-1	8.71
Stinson 108	10.65	Navion	7.84
Ercoupe	9.50	Piper J-3	7.61

Aircraft	Rate	Aircraft	Rate
Luscombe 8	7.58	Mooney M-20	3.42
Cessna 120/140	6.73	Piper PA-18	3.37
Piper PA-12	6.54	Cessna 177	3.33
Bellanca 14-19	5.98	Cessna 206	3.30
Piper PA-22	5.67	Cessna 180	3.24
Cessna 195	4.69	Cessna 170	2.88
Piper PA-32	4.39	Cessna 185	2.73
Cessna 210/205	4.25	Cessna 150	2.48
Aeronca 7	4.23	Piper PA-28	2.37
Aeronca 11	4.10	Beech 33,35,36	2.22
Taylorcraft BC	3.81	Grumman AA-5	2.20
Piper PA-24	3.61	Cessna 182	2.08
Beech 23	3.58	Cessna 172	1.41
Cessna 175	3.48		

In-Flight Airframe Failure

Aircraft	Rate	Aircraft	Rate
Bellanca 14-19	1.49	Beech 23	0.27
Globe GC-1	1.03	Cessna 120/140	0.27
Ercoupe	0.97	Piper PA-32	0.24
Cessna 195	0.94	Taylorcraft BC	0.24
Navion	0.90	Piper J-3	0.23
Aeronca 11	0.59	Mooney M-20	0.18
Beech 33,35,36	0.58	Piper PA-28	0.16
Luscombe 8	0.54	Cessna 177	0.16
Piper PA-24	0.42	Cessna 182	0.12
Cessna 170	0.36	Cessna 206	0.11
Cessna 210/205	0.34	Grumman AA-1	0.09
Cessna 180	0.31	Cessna 172	0.03
Piper PA-22	0.30	Cessna 150	0.02
Aeronca 7	0.27		

Stall

Aircraft	Rate	Aircraft	Rate
Aeronca 7	22.47	Cessna 170	4.38
Aeronca 11	8.21	Grumman AA-1	4.23
Taylorcraft BC	6.44	Piper PA-12	3.27
Piper J-3	5.88	Cessna 120/140	2.51
Luscombe 8	5.78	Stinson 108	2.09
Piper PA-18	5.49	Navion	1.81
Globe GC-1	5.15	Piper PA-22	1.78

Aircraft	Rate	Aircraft	Rate
Cessna 177	1.77	Piper PA-28	0.80
Grumman AA-5	1.76	Mooney M-20	0.80
Cessna 185	1.47	Cessna 172	0.77
Cessna 150	1.42	Cessna 210/205	0.71
Beech 23	1.41	Bellanca 14-19	0.60
Ercoupe	1.29	Piper PA-32	0.57
Cessna 180	1.08	Cessna 206	0.54
Piper PA-24	0.98	Cessna 195	0.47
Beech 33,35,36	0.94	Cessna 182	0.36
Cessna 175	0.83		

Hard Landing

Aircraft	Rate	Aircraft	Rate
Beech 23	3.50	Cessna 175	1.00
Grumman AA-1	3.02	Cessna 180	0.93
Ercoupe	2.90	Cessna 210/205	0.82
Cessna 177	2.60	Piper PA-28	0.81
Globe GC-1	2.58	Cessna 172	0.71
Luscombe 8	2.35	Piper PA-22	0.69
Cessna 182	2.17	Taylorcraft BC	0.48
Cessna 170	1.89	Cessna 195	0.47
Beech 33,35,36	1.45	Piper PA-18	0.43
Cessna 150	1.37	Piper PA-32	0.42
Cessna 120/140	1.35	Cessna 185	0.42
Cessna 206	1.30	Navion	0.36
Piper PA-24	1.29	Mooney M-20	0.31
Aeronca 7	1.20	Piper PA-12	0.23
Piper J-3	1.04	Stinson 108	0.19
Grumman AA-5	1.03		

Ground Loop

Aircraft	Rate	Aircraft	Rate
Cessna 195	22.06	Piper PA-12	4.67
Stinson 108	13.50	Piper PA-18	3.90
Luscombe 8	13.00	Taylorcraft BC	3.58
Cessna 170	9.91	Globe GC-1	3.09
Cessna 120/140	8.99	Grumman AA-1	2.85
Aeronca 11	7.86	Piper PA-22	2.76
Aeronca 7	7.48	Ercoupe	2.74
Cessna 180	6.49	Beech 23	2.33
Cessna 185	4.72	Bellanca 14-19	2.10

Aircraft	Rate	Aircraft	Rate
Piper J-3	2.07	Cessna 210/205	1.08
Cessna 206	1.73	Cessna 182	1.06
Cessna 177	1.61	Cessna 172	1.00
Grumman AA-5	1.47	Mooney M-20	0.65
Piper PA-32	1.42	Beech 33,35,36	0.55
Cessna 150	1.37	Navion	0.36
Piper PA-28	1.36	Cessna 175	0.17
Piper PA-24	1.29		

Undershoot

Aircraft	Rate	Aircraft	Rate
Ercoupe	2.41	Cessna 195	0.47
Luscombe 8	1.62	Grumman AA-5	0.44
Piper PA-12	1.40	Piper PA-18	0.43
Globe GC-1	1.03	Beech 23	0.43
Cessna 175	0.99	Cessna 185	0.41
Grumman AA-1	0.95	Mooney M-20	0.37
Taylorcraft BC	0.95	Cessna 170	0.36
Piper PA-22	0.83	Navion	0.36
Piper PA-32	0.70	Cessna 150	0.35
Bellanca 14-19	0.60	Cessna 210/205	0.33
Aeronca 11	0.59	Cessna 206	0.32
Piper PA-28	0.59	Cessna 172	0.26
Aeronca 7	0.59	Cessna 182	0.24
Piper PA-24	0.57	Beech 33,35,36	0.21
Piper J-3	0.57	Cessna 180	0.15
Stinson 108	0.57	Cessna 177	0.10
Cessna 120/140	0.53		

Overshoot

Aircraft	Rate	Aircraft	Rate
Grumman AA-5	2.35	Luscombe 8	1.08
Cessna 195	2.34	Piper PA-32	1.03
Beech 23	1.95	Globe GC-1	1.03
Piper PA-24	1.61	Mooney M-20	1.01
Piper PA-22	1.33	Cessna 172	1.00
Cessna 175	1.33	Cessna 170	0.99
Stinson 108	1.33	Grumman AA-1	0.95
Cessna 182	1.21	Piper PA-12	0.93
Aeronca 11	1.17	Cessna 210/205	0.89

Aircraft	Rate	Aircraft	Rate
Cessna 177	0.88	Cessna 180	0.56
Piper PA-18	0.81	Navion	0.54
Cessna 206	0.81	Aeronca 7	0.48
Piper PA-28	0.80	Cessna 150	0.35
Cessna 120/140	0.71	Piper J-3	0.34
Ercoupe	0.64	Cessna 185	0.31
Bellanca 14-19	0.60	Beech 33,35,36	0.23

Owners Say:

"Maintenance is what I like about my Cherokee. It requires very little, and I do most of it—under supervision, of course."

"I wish my Warrior had a left door."

"Stalls are almost nonexistent. Just a little up-down, and a slight shudder."

"I have a Cherokee 180 with the fat wings. You have to flare correctly on landings; there's almost no ground effect."

"About six years ago I sold our [Bellanca] Viking. It had gotten too expensive to own. The maintenance was drowning me. I figured I was done in the airplane ownership business. Then the bug bit and I went for a new biennial in a Warrior II. Now I own one. It may be a little slower, but it still is a very classy bird, and I can afford to keep it up."

"I used to have an old 172 with manual flaps. My late model Warrior has them too. That's good!"

"My last annual cost $345, and that's in the high-cost Washington, D.C., area. I'm real happy with that."

"Several years ago I owned a Cessna Skyhawk with ARC radios. Those radios never worked. When I got this Warrior, it had all Narco equipment. I've been very pleased with it for the past three years."

"My Six is like a station wagon. We have two kids, and we take it all over the country. We've never had more baggage than space."

"I like the openness of the low-wing airplane—better vision too."

"I like the way my Archer handles in high winds. Loads easier than my old [Cessna] 172."

"I could afford to buy a retractable if I wanted it, but why pay more in extra maintenance for a little extra speed? Slow down and enjoy life . . . fly a Warrior!"

Salespeople Say:

"If a person comes to me looking for a two-place, I always try to get him into an older PA-28. After you sit in one of them, you'll never want to get into another Cessna 150. It's so much bigger. Besides, the odds are that one day there'll be a need for the extra seats.''

"The Warriors sell themselves. They're roomy, look good, and are reasonable in price. They're just real good value . . . something you don't often see these days.''

"The Arrow makes a real fine family airplane—nice size to handle, yet very fast.''

"I've sold several [PA-32] Sixes to small charter outfits that use them off-airport for the well-drilling companies. They seem to be very tough, 'cause those folks don't spend money easily.''

"My personal airplane is an Archer II. It has a plush interior, curtains at the windows, and flower glasses on the sides (always with a rose in each). When a man and wife come looking at airplanes, I always make sure the wife sees mine. After that, the scales are tipped in favor of the low-wing planes.''

TIPS

Never leave your airplane unattended and not tied down. It could not only move in the wind and get damaged, it could damage someone else's airplane, and you would be hit with *all* the repair bills.

Put plugs in the air intakes of the cowling to keep the birds out of the engine compartment.

Use pitot tube covers to keep bugs from blocking the little holes.

Mount a fire extinguisher in the cabin where you can *quickly* get at it.

Keep a working flashlight on board.

Always carry a complete first-aid kit in your plane.

If you are flying over sparsely populated areas, it would be a good idea to have survival water, food, and cover (blankets or sleeping bags) on board.

Having flares on board can be handy if you are down and trying to attract the attention of a rescue team.

TEST YOURSELF

The following quiz is designed to make you more aware of your airplane's characteristics and limitations. Perhaps you'll have to do a little checking up on your numbers to answer the questions. There is no pass or fail grade; only you know if you did well.

For *your* aircraft:

1. What is the normal climb speed?
2. What is the best-rate-of-climb speed?
3. What is the best-angle-of-climb speed?
4. What is the maximum flaps-down speed?
5. What is the maximum gear-down speed?
6. What is the "clean" stall speed?
7. What is the "dirty" stall speed?
8. What is the approach speed?
9. What is the maneuvering speed?
10. What is the never-exceed speed?
11. What glide speed will give you maximum range?
12. How many usable gallons of fuel can your plane carry?
13. Describe how you drain the fuel sumps.
14. At 65-percent power at 5,000 feet, what is your maximum endurance (in hours and minutes)?
15. How do you figure estimated true airspeed at 5,000 feet?
16. What rpm or rpm/manifold pressure gives 65 percent power at 5,000 feet?
17. How many gallons of fuel are used per hour at 65 percent power?
18. What is the make and hp of your engine?
19. What fuel octane rating does your engine require?
20. What make, type, and weight oil do you use?
21. What is the maximum takeoff weight for your plane?
22. Describe how the maximum takeoff weight is affected by density altitude.
23. What is the maximum crosswind component for your plane?
24. What are the common UNICOM frequencies?
25. What is the COMM emergency frequency?

Appendix A

Typical Annual Inspection

The FAA, in its desire to fill the airways with only airworthy aircraft, requires that each aircraft be inspected for proper operation and structural integrity on a continuing basis. For the average owner this means annually, i.e., the annual inspection.

The following are the items generally inspected during an annual inspection. There is no standard FAA inspection list, only the requirement for a "complete inspection."

The following list contains recommended items to be checked. Details as to how to check, or what to check for, are not included. The inspection list is general in nature, and may be applied to most small general aviation aircraft. Some items may apply only to specific models, and some items apply to optional equipment that may or may not be found on a particular airplane.

Paperwork Check

Miscellaneous data, information, and licenses are a part of the airplane file. Check that the following documents are up-to-date, in accordance with current Federal Aviation Regulations, and displayed in the airplane at all times:

- ☐ Airworthiness Certificate
- ☐ Registration Certificate
- ☐ Aircraft Radio License

To be carried in the airplane at all times:

- ☐ Weight-and-Balance and associated papers
- ☐ Aircraft Equipment List

To be made available upon request:

☐ Aircraft Logbook

☐ Engine Logbook

A complete check for AD compliance is a part of an annual inspection.

Mechanical Check

Inspect all movable parts for:

☐ Lubrication   ☐ Cracked fittings
☐ Security of attachment   ☐ Security of hinges
☐ Binding   ☐ Defective bearings
☐ Excessive wear   ☐ Cleanliness
☐ Safetying   ☐ Corrosion
☐ Proper operation   ☐ Deformation
☐ Proper adjustment   ☐ Sealing
☐ Correct travel   ☐ Tensions

Inspect non-movable metal parts for:

☐ Security of attachment   ☐ Corrosion
☐ Cracks   ☐ Condition of paint
☐ Metal distortion   ☐ Other apparent damage
☐ Broken spot welds

Inspect fluid lines and hoses for:

☐ Leaks   ☐ Kinks
☐ Cracks   ☐ Chafing
☐ Dents   ☐ Proper radius
☐ Security   ☐ Obstructions
☐ Corrosion   ☐ Foreign matter
☐ Deterioration

Inspect wiring for:

☐ Security   ☐ Loose or broken terminals
☐ Chafing   ☐ Heat deterioration
☐ Burning   ☐ Corroded terminals
☐ Defective insulation

Check bolts/screws for:

☐ Correct torque, as installed or when visual inspection indicates the need for a torque check.

Inspect filters, screens, and fluids for:

☐ Cleanliness
☐ Contamination
☐ Replacement at specified intervals

Engine Check

Start, run up, and shut down the engine(s) in accordance with instructions in the owner's manual. During the run-up, observe the following (make note of any discrepancies or abnormalities):

☐ Engine temperatures or pressures
☐ Static rpm
☐ Magneto drop
☐ Engine response to changes in power
☐ Any unusual engine noises
☐ Fuel selector valve (operate the engine[s] on each position long enough to make sure the valve functions properly)
☐ Idling speed and mixture
☐ Proper idle cutoff
☐ Alternator
☐ Suction gauge

Note: After completion of the inspection, the engine run-up should again be performed to ascertain that any discrepancies or abnormalities have been corrected.

Propeller Check

☐ Spinner and spinner bulkhead
☐ Blades
☐ Hub
☐ Bolts and/or nuts

Engine Compartment Check

Check for evidence of oil and fuel leaks, then clean the entire engine and compartment, if needed, prior to inspection.

☐ Engine oil, screen, filler cap, dipstick, drain plug, and filter
☐ Oil cooler

- [ ] Induction air filter element
- [ ] Induction airbox, air valves, doors, and controls
- [ ] Engine baffles
- [ ] Cylinders, rocker box covers, and pushrod housings
- [ ] Crankcase, oil sump, accessory section, and front crankshaft seal
- [ ] All lines and hoses
- [ ] Vacuum pump, oil separator, and relief valve
- [ ] Vacuum relief valve filter
- [ ] Engine controls and linkage
- [ ] Engine shock mounts, engine mount structure, and ground straps
- [ ] Cabin heater valves, doors, and controls
- [ ] Electrical wiring
- [ ] Starter, solenoid, and electrical connections brushes, brush leads, and commutator
- [ ] Alternator, and electrical connections brushes, brush leads, and commutator or slip ring
- [ ] Voltage regulator mounting and electrical leads
- [ ] Magnetos, electrical connections, and timing
- [ ] Intake and exhaust systems
- [ ] Ignition harness
- [ ] Spark plugs and compression
- [ ] Carburetor
- [ ] Fuel injection system
- [ ] Turbocharger
- [ ] Firewall

Fuel System Check

- [ ] Fuel strainer, drain valve control, strainer screen, and bowl
- [ ] Fuel tanks, fuel lines, drains, filler caps, and placards
- [ ] Drain fuel and check tank interior, attachment, and outlet screens
- [ ] Fuel vents and vent valves
- [ ] Fuel selector valve and placards
- [ ] Engine primer

Landing Gear Check

- [ ] Brake fluid, lines and hoses, linings, discs, brake assemblies, and master cylinders

- [ ] Wheels, wheel bearings, strut, tires, and fairings
- [ ] Wheel bearing lubrication
- [ ] Torque link lubrication
- [ ] Gear strut servicing
- [ ] Parking brake system
- [ ] Retraction and extension systems

Airframe Check

- [ ] Aircraft exterior/interior structure
- [ ] Windows, windshield, and doors
- [ ] Skin and structure of control surface and trim tabs
- [ ] Balance weight attachment
- [ ] Seats, stops, seat rails, upholstery, structure, and seat mounting
- [ ] Safety belts and attaching brackets
- [ ] Instruments and markings
- [ ] Magnetic compass compensation
- [ ] Instrument wiring and plumbing
- [ ] Instrument panel, shock mounts, ground straps, cover, and decals and labeling
- [ ] Defrosting, heating, and ventilating systems and controls
- [ ] Cabin upholstery, trim, sun visors, and ashtrays
- [ ] Area beneath floor, lines, hoses, wires, and control cables
- [ ] Lights, switches, circuit breakers, fuses, and spare fuses
- [ ] Exterior lights
- [ ] Pitot and static system
- [ ] Stall warning sensing unit, pitot heater
- [ ] Radios and antennas
- [ ] Battery, battery box, battery cables and electrolyte level

Control Systems Check

Always check for correct direction of movement, correct travel, and correct cable tension.

- [ ] Cables, terminals, pulleys, pulley brackets, cable guards, turnbuckles, and fairleads
- [ ] Chains, terminals, sprockets, and chain guards
- [ ] Control bearings, sprockets, pulleys, cables, chains, and turnbuckles, control lock, wheel

- ☐ Trim control wheels, indicators, and actuators
- ☐ Travel stops
- ☐ All decals and labeling
- ☐ Flap controls

Appendix B
Landing Gear
Troubleshooting Chart

Nosewheel shimmies during high-speed operations (i.e., takeoff, landing, high-speed taxi):
- ☐ Shimmy damper worn
- ☐ Shimmy damper loose at mounting
- ☐ Tire out-of-balance
- ☐ Worn/loose wheel bearings
- ☐ Worn link bolts or bushings

Unusual/excessive nose tire wear:
- ☐ Incorrect air pressure
- ☐ Shimmy problems

Poor steering:
- ☐ Oleo strut binding
- ☐ Dragging brakes on main wheels
- ☐ Loose bellcrank mount
- ☐ Shimmy damper binding

Main wheel(s) shimmy:
- ☐ Tire(s) out of balance
- ☐ Worn/loose wheel bearings
- ☐ Worn torque link bolts/bushings

Unusual/excessive main tire wear:
- ☐ Incorrect air pressure
- ☐ Incorrect alignment

Strut(s) bottom on normal landing:
- ☐ Needs air/fluid in strut

Indicator lights operating incorrectly:
- ☐ Lamp burned out
- ☐ Wiring difficulties
- ☐ Circuit breaker open
- ☐ Limit switches out of adjustment
- ☐ Gear not retracting properly

Warning horn inoperative:
- ☐ Circuit wire broken
- ☐ Defective safety switch

Warning light/horn fails to shut off:
- ☐ Throttle microswitch out of adjustment

Landing gear doors fail to close completely:
- ☐ Gear not retracting completely
- ☐ Door retraction rods need adjustment

Appendix C

Engine Troubleshooting Chart

Engine fails to start:
- ☐ Lack of fuel
- ☐ Underpriming
- ☐ Flooding
- ☐ Incorrect throttle setting
- ☐ Defective spark plug(s)
- ☐ Defective ignition wire(s)
- ☐ Magneto timing or breaker points
- ☐ Defective magneto wiring
- ☐ Mixture control at idle cutoff

Engine won't idle:
- ☐ Carburetor needs adjustment
- ☐ Air leak in the intake manifold
- ☐ Low cylinder compression
- ☐ Faulty ignition system
- ☐ Dirty air filter element

Rough engine (low rpm):
- ☐ Fuel/air mixture too rich
- ☐ Fuel/air mixture too lean
- ☐ Air leaks in intake manifold
- ☐ Faulty ignition system
- ☐ Restricted exhaust system

Rough engine (high rpm):
- ☐ Unbalanced propeller
- ☐ Damaged propeller
- ☐ Defective engine mounts
- ☐ Fouled spark plugs

Low power output:
- ☐ Throttle control out of adjustment
- ☐ Air leak in the intake manifold
- ☐ Restricted air intake
- ☐ Improper grade/type fuel
- ☐ Faulty ignition system
- ☐ Insufficient fuel flow

Low oil pressure:
- ☐ Needs oil
- ☐ Dirty oil screen
- ☐ Defective gauge
- ☐ Clogged relief valve
- ☐ High oil temperature
- ☐ Damaged engine bearings

High oil pressure:
- ☐ Excessive blow-by
- ☐ Defective bearings
- ☐ Clogged oil lines/screens
- ☐ Needs oil

Excessive oil consumption:
- ☐ Defective bearings
- ☐ Worn/broken piston rings
- ☐ Oil leakage (gaskets)

Appendix D

Fuel System

Troubleshooting Chart

No fuel flow:
- ☐ Fuel line blocked
- ☐ Vent cap clogged
- ☐ Defective fuel pump
- ☐ Defective fuel selector valve

Fuel gauge inoperative:
- ☐ No fuel in tanks
- ☐ Broken circuit wire
- ☐ Defective sending unit in tank
- ☐ Defective gauge
- ☐ Open circuit breaker

Fuel gauge reads full all the time:
- ☐ Broken ground wire at sending unit

No fuel pressure indicated:
- ☐ Defective gauge
- ☐ Defective fuel pump
- ☐ Fuel valve stuck
- ☐ No fuel in tanks

Low/surging fuel pressure:
- ☐ Obstruction at inlet side of fuel pump
- ☐ Defective bypass valve
- ☐ Faulty fuel pump

Leaking selector valve:
- ☐ Worn O-ring seals

Appendix E
Battery Troubleshooting Chart

Hydrometer Reading/Battery Charge Percent:

Reading	Percent of charge
1280	100
1250	75
1220	50
1190	25
1160	Little useful capacity
1130 (or below)	Discharged

Discharged battery:
- ☐ Worn-out battery
- ☐ Charging rate too low
- ☐ Short circuit in wiring
- ☐ Equipment left on

Short battery life:
- ☐ Needs electrolyte (water)
- ☐ Charging rate too low
- ☐ Battery is worn out

Frozen battery:
- ☐ Thaw and discard (A battery kept at full charge will not freeze. In other words, take care of it!)

Excessive corrosion in box:
- ☐ Clean and flush area (use baking soda and water for cleaning)

Consumes excessive electrolyte:
- ☐ Charging rate too high

Index